THE REVOLUTION OF AI IN MEDICINE

AI Revolution in Medicine, The Future of Healthcare

REYNOLD JAMESON

Copyright © 2023 by Reynolds Jameson

All rights reserved. No part of this publication may be reproduced, distributed, or transmitted in any form or by any means, including photocopying, recording, or other electronic or mechanical methods, without the prior written permission of the publisher, except in the case of brief quotations embodied in critical reviews and certain other noncommercial uses permitted by copyright law.

For permissions requests, write to the publisher.

Disclaimer: The facts and views contained in this book are meant for informative purposes only. The author and publisher make no claims or guarantees with regard to the accuracy or completeness of the contents of this book and particularly disclaim any implied warranties of merchantability or suitability for a particular purpose. The material provided in this book is not meant to substitute professional medical advice or treatment. Readers are advised to check with healthcare specialists for any medical issues or inquiries they may have.

TABLE OF CONTENTS

copyright 0
Table of Contents 2
Introduction 5
Chapter 1: The Rise of AI in Healthcare 7
1.1 The Emergence of AI in Healthcare 7
1.2 Understanding AI Technologies 12
1.3 Early Applications of AI in Medicine 17
Chapter 2: AI and Diagnostics 25
2.1 The Importance of Accurate Diagnostics 25
2.2 AI in Diagnostic Imaging 29
2.3 AI in Pathology and Laboratory Medicine 37
2.4 AI in Diagnostic Decision Support 46
2.5 Challenges and Ethical Considerations in AI Diagnostics . 55
2.6 Future Perspectives and Impact of AI on Diagnostics 63
Chapter 3: AI in Treatment and Personalized Medicine 71
3.1 The Paradigm Shift towards Personalized Medicine 71
3.3 Precision Treatment Selection with AI 81
3.4 AI-Assisted Treatment Planning and Decision-Making 88
3.5 Real-Time Monitoring and Adaptive Treatment 94
3.6 Ethical Considerations in AI-Driven Treatment 102
3.7 Regulatory and Adoption Challenges in AI Treatment 108
3.8 Future Perspectives and Impact of AI in Personalized Medicine 115

Chapter 4: AI in Medical Imaging 118

4.1 Introduction to Medical Imaging 118

4.2 Evolution of AI in Medical Imaging................................. 121

4.3 AI in Nuclear Medicine and Molecular Imaging.............. 124

Chapter 5: AI and Electronic Health Records........................ 129

5.1 Introduction to Electronic Health Records (EHR)............ 129

5.2 AI Applications in EHR.. 133

5.3 Natural Language Processing (NLP) in EHR 138

5.4 AI-Enabled Clinical Workflow Optimization................... 144

5.5 AI and Interoperability of EHR Systems.......................... 151

5.6 Enhancing EHR Data Quality with AI 157

5.7 Ethical Considerations in AI-Driven EHR 163

5.8 Legal and Regulatory Challenges in AI-Driven EHR 169

Chapter 6: AI and Patient Care 175

6.1 Transforming Patient Care with AI 175

6.2 AI in Patient Monitoring and Management 177

6.3 AI in Medication Management and Adherence................ 180

6.4 AI in Patient Engagement and Education 184

6.5 AI in Mental Health and Well-being 188

6.6 AI-Enabled Precision Rehabilitation 190

6.7 AI in End-of-Life Care and Palliative Care 193

Chapter 7: Ethical and Legal Considerations......................... 198

7.1 Introduction to Ethical and Legal Considerations in AI ... 198

7.2 Ethical Principles in AI Development and Deployment... 199

7.4 Legal Frameworks for AI Regulation 211

7.5 Ethical Considerations in AI Applications 218
7.6 Dealing with Discrimination and Bias in AI Systems 227
7.7 Moral Issues with AI Decision-Making........................... 233
7.8 AI Ethics in Research and Development 239
7.9 Ensuring Ethical and Legal Compliance in AI 247
Conclusion 253

INTRODUCTION

The area of healthcare is continually developing, driven by developments in technology and the rising desire for more efficient and effective patient care. In recent years, Artificial Intelligence (AI) has emerged as a strong technology with the potential to change healthcare and redefine the way medical treatments are done. AI provides the possibility of enhancing diagnoses, therapy selection, patient monitoring, and overall healthcare delivery.

This book provides as a complete guide to the implementation of AI in healthcare. It addresses the numerous fields within healthcare where AI is having a big influence, including customized medicine, medical imaging, electronic health records, patient care, and ethical implications in AI-driven healthcare. The chapters discuss the principles of AI and its integration into diverse parts of healthcare, offering insights into the possible advantages, problems, and ethical concerns.

Chapter 1 presents the fundamental ideas of AI and its relation to healthcare. It establishes the groundwork for comprehending the upcoming chapters by discussing fundamental vocabulary, AI methodologies, and the function of machine learning algorithms in healthcare applications.

Chapter 2 goes into the notion of customized medicine and how AI is creating a paradigm change in the sector. It highlights the knowledge of individual patients' features, genetic data, and therapy response prediction, paving the way for precision medicine.

Chapter 3 focuses on AI in medication development and therapy selection. It analyzes how AI is expediting the drug development process, assisting in target identification and validation, and enabling medication repurposing and combination therapy.

AI Revolution in Medicine, The Future of Healthcare

Chapter 4 analyzes the role of AI in medical imaging, where it is changing modalities such as ultrasound, computed tomography (CT), magnetic resonance imaging (MRI), nuclear medicine, and molecular imaging. It covers the breakthroughs in AI algorithms for image analysis, segmentation, measuring, and lesion diagnosis.

Chapter 5 investigates the incorporation of AI in electronic health records (EHR) systems. It examines the advantages and problems of EHR deployment, AI applications in data mining, predictive analytics, clinical decision support systems, and natural language processing for text mining and information extraction.

Chapter 6 focuses on AI in patient care, covering issues such as remote patient monitoring, predictive analytics for illness identification, medication management, patient engagement, mental health, and end-of-life care. It illustrates how AI is increasing patient outcomes and improving the overall quality of care.

Chapter 7 examines the ethical and legal concerns related with AI in healthcare. It tackles ethical principles in AI development, fairness and bias reduction, openness and explainability, privacy and data protection, accountability and responsibility, legal frameworks, and ethical difficulties in AI decision-making.

This book gives a complete overview of the use of AI in healthcare. It covers a broad variety of issues, including personalized medicine, medical imaging, electronic health records, patient care, and ethical concerns. By addressing the advantages, problems, and ethical implications, readers will acquire vital insights into the revolutionary potential of AI in healthcare and the significance of responsible and ethical use of these technologies.

AI Revolution in Medicine, The Future of Healthcare

CHAPTER 1: THE RISE OF AI IN HEALTHCARE

1.1 The Emergence of AI in Healthcare

1.1.1 A Brief History of AI in Medicine

In order to understand the current shift of AI in healthcare, it is important to follow the beginnings and development of AI within the field of medicine. While the roots of AI can be traced back to the mid-20th century, its application in medicine gained popularity in more recent decades.

The seeds of AI in medicine were sown in the 1950s when early computer scientists and researchers began studying the potential of machines to simulate human intelligence. During this time, experts such as Warren McCulloch and Walter Pitts built the basis for neural network theory, a basic idea driving many AI uses today.

In the 1970s and 1980s, the area of medical expert systems emerged. Expert systems tried to collect and emulate the information and decision-making processes of human experts. The creation of programs like MYCIN, an expert system for detecting bacterial illnesses, and CADUCEUS, a system for diagnosing and controlling specific diseases, showed the early promise of AI in healthcare.

Advancements in computer power and data access in the 1990s drove the growth of AI in medicine. Machine learning algorithms, which allow systems to learn trends and make predictions from data, gained popularity. The application of machine learning methods to medical data allows for the development of prediction models and decision support tools.

AI Revolution in Medicine, The Future of Healthcare

In the early 2000s, the merging of AI technologies with medical images began to change the field. Computer vision techniques allowed the automatic study and understanding of medical pictures, boosting the skills of doctors and pathologists. This marked a significant change in the evaluation and treatment planning processes, where AI systems could help in spotting problems and directing medical experts.

In more recent years, the rise of big data and the availability of electronic health records (EHRs) have driven the growth of AI in healthcare. The wealth of medical data, combined with advanced analytics and deep learning methods, has allowed the development of complex models for disease forecast, individual treatment suggestions, and community health management.

Today, AI in medicine is at the center of innovation, with a myriad of uses across various fields. From identifying diseases to helping in medical processes, AI is changing healthcare service, study, and patient results.

As we journey further into the change of AI in medicine, it is important to think on the rich past and development of AI in healthcare. By building upon the knowledge and successes of the past, we can tap the full potential of AI to change patient care and pave the way for a future where AI and human experience jointly advance the limits of medicine.

1.1.2 Technological Advancements Driving the Rise of AI

The fast rise of AI in healthcare can be credited to several key technological breakthroughs that have provided a healthy ground for its growth and integration within the medical field. These developments have expanded the possibilities of AI applications, driven by the ever-increasing computing power, data access, and breakthroughs in algorithms. Let's study some of the technology causes that have pushed the rise of AI in healthcare.

AI Revolution in Medicine, The Future of Healthcare

1. Increased computer Power: The exponential growth in computer power over the past few decades has been crucial in improving AI in healthcare. More powerful computers, graphics processing units (GPUs), and specialized hardware accelerators have allowed the fast handling of complex AI algorithms. This improved processing power has allowed the training and application of deep learning models that can handle vast amounts of medical data and extract useful insights.

2. Big Data and Electronic Health Records (EHRs): The digital change in healthcare has led to the gathering of vast amounts of data, including electronic health records, medical imaging data, genetic information, and smart device data. This wealth of information, generally referred to as big data, has given a rich supply for training AI models. EHRs, in particular, have played a key role in supporting AI breakthroughs by offering complete patient data that can be harnessed to create prediction models, decision support systems, and personalized treatment suggestions.

3. advances in Algorithms: Significant advances in AI algorithms, especially in the field of machine learning and deep learning, have been important in the rise of AI in healthcare. Machine learning methods such as guided learning, unsupervised learning, and reinforcement learning have allowed the development of predictive models, grouping algorithms, and anomaly detection systems. Deep learning, driven by neural networks with multiple layers, has transformed picture analysis, natural language processing, and pattern recognition tasks. These programs have the ability to learn from big datasets and make correct guesses or help in difficult decision-making processes.

4. Image and Signal Processing: AI has made great progress in the study of medical photos and data. Image processing methods, mixed with AI algorithms, have allowed the automatic analysis of radiological images, tissue slides, and other diagnostic imaging

tools. AI-powered picture analysis can help in spotting abnormalities, segmenting structures, and aiding doctors and scientists in their decision-making processes. Similarly, AI algorithms have found uses in the study of bodily data, such as electrocardiograms (ECGs), electroencephalograms (EEGs), and vital signs tracking, enabling early spotting of problems and better patient care.

5. Natural Language Processing (NLP): Advancements in natural language processing have allowed AI systems to understand and pull meaning from written data, including physician notes, study books, and patient-generated material. NLP methods, mixed with machine learning algorithms, allow the automation of jobs such as clinical recording, information search, and mood analysis. AI-powered virtual helpers and robots equipped with NLP skills have improved patient interaction, simplified routine processes, and facilitated the delivery of personalized healthcare information.

These technological developments have mutually fueled the rise of AI in healthcare, enabling doctors, researchers, and healthcare organizations to leverage the power of intelligent technologies in changing patient care, medical research, and decision-making processes. As the pace of technological progress continues, the potential for further developments and breakthroughs in AI uses within medicine is endless.

1.1.3 The Convergence of Medicine and Technology

The rise of AI in healthcare indicates a deep merging between the areas of medicine and technology. This merger has created a strong synergy that holds huge potential to change patient care, better medical study, and shape the future of healthcare. Let's discuss the key aspects of this merger and the changing effect it has on the practice of medicine.

1. breakthroughs in Medical Technology: Medical technology has made great steps in recent years, driven by breakthroughs in imaging systems, smart devices, genetic sequencing, and robots, among others. These technological advances have not only improved diagnosis skills but have also created a huge amount of data, which acts as the fuel for AI algorithms. AI has emerged as a powerful tool to analyze, understand, and draw insights from this huge amount of medical data, thereby improving the powers of healthcare workers and allowing more accurate diagnoses, personalized treatments, and proactive interventions.

2. Data-Driven Healthcare: The merging of medicine and technology has led to a paradigm move towards data-driven healthcare. Electronic health records, genetic data, medical images, real-time tracking devices, and patient-generated data have become essential components of the modern healthcare environment. AI programs can process and analyze these varied datasets, finding trends, predicting outcomes, and allowing evidence-based decision-making. By tapping the power of AI, healthcare workers can gain useful insights from patient data, leading to more accurate evaluations, better treatment plans, and improved patient results.

3. Precision Medicine: One of the most transformative parts of the merger between medicine and technology is the rise of precision medicine. By mixing DNA information, clinical data, and AI-powered analyses, precision medicine aims to offer targeted healthcare tuned to an individual's unique genetic makeup, lifestyle, and external factors. AI systems play a crucial role in understanding the complex links between genetic variants, disease risk, and treatment reactions. This personalized method allows healthcare workers to improve medicines, reduce side effects, and increase treatment effectiveness, eventually leading to better patient results.

4. Augmented Decision-Making: The merging of AI technologies within healthcare has the ability to assist and increase

the decision-making process of healthcare workers. AI programs can handle huge amounts of medical literature, clinical standards, and study results to provide evidence-based suggestions and help doctors in making informed choices. By leveraging machine learning and natural language processing, AI systems can quickly analyze and summarize relevant information, allowing healthcare workers to stay up-to-date with the latest advancements and make well-informed choices for their patients.

5. Patient-Centric Care: The merging of medicine and technology puts a larger focus on patient-centric care. AI-powered technologies, such as virtual helpers and chatbots, allow improved patient interaction, preventive health tracking, and personalized health information delivery. These tools enable patients to take an active part in their healthcare path, promoting a joint and informed approach to healthcare delivery. Moreover, AI-driven prediction analytics can spot people at higher risk of having certain conditions, allowing early treatments and preventive steps, thereby promoting general well-being.

The merging of medicine and technology through AI signifies a radical change in healthcare. It holds the potential to change the way healthcare is provided, empower people, increase medical study, and improve health results. As this convergence deepens, it is crucial to navigate the ethical, regulatory, and societal implications to ensure the responsible and equitable integration of AI technologies into the fabric of healthcare, keeping the well-being of patients and the sanctity of the physician-patient relationship at the forefront.

1.2 Understanding AI Technologies

1.2.1 Machine Learning: Foundation of AI in Healthcare

Machine learning is a basic component of AI that serves as the basis for many uses in healthcare. It is a form of AI that allows computers to learn from data and make guesses or take actions without being explicitly coded. In the setting of healthcare, machine learning methods have the ability to change medical study, analysis, treatment, and patient care. Let's dig into the key aspects of machine learning and its uses in healthcare.

1. Machine Learning Basics: Machine learning algorithms are meant to spot patterns, connections, and trends within datasets. These algorithms learn from past data by pulling traits and building mathematical models that can be used to predict results or make decisions. The learning process includes teaching the algorithm on a named dataset, allowing it to repeatedly improve its performance through feedback and changes.

2. Supervised Learning: Supervised learning is a type of machine learning in which the program learns from named examples. The program is given with a collection where each case is linked with a known result or goal variable. By studying the input traits and related names, the program learns to make guesses or describe new, unknown data correctly. In healthcare, supervised learning can be applied to tasks such as disease evaluation, predicting treatment reactions, and risk assessment.

3. Unsupervised Learning: In comparison to supervised learning, unsupervised learning methods aim to find hidden patterns or structures within unidentified data. These programs evaluate the data without prior knowledge of the results or labels. Unsupervised learning methods such as grouping and dimensionality reduction can be used to discover connections in complicated datasets, spot patient subgroups, or gain insights into disease development.

4. Reinforcement Learning: Reinforcement learning involves an animal learning to make choices through trial and error exchanges with an environment. The agent gets input in the form of rewards or punishments based on its actions, allowing it to learn best methods for maximizing benefits over time. Reinforcement learning has the

potential to be applied in healthcare situations, such as improving treatment plans or resource sharing, by constantly changing tactics based on the input received from the environment.

5. Applications of Machine Learning in Healthcare: Machine learning methods have a wide range of applications in healthcare, covering different areas. Some famous examples include:

• Disease identification: Machine learning models can examine patient data, such as signs, medical images, test results, and genetic information, to help in disease identification. These models can help healthcare workers in making accurate and quick decisions, especially for complicated illnesses or rare diseases.

• Predictive Analytics: Machine learning systems can study patient data to predict disease development, treatment results, or possible bad events. These prediction models can allow preventative actions, optimize treatment plans, and improve patient management and care coordination.

• Drug development and Development: Machine learning methods can be applied in drug development processes to speed the finding of possible medicinal compounds, improve drug formulas, and predict drug-target interactions. By leveraging large-scale biological and chemical information, machine learning methods aid in the finding and development of new drugs.

• Healthcare Operations and Resource Management: Machine learning methods can improve hospital operations by predicting patient flow, expecting demand for resources, and allowing efficient schedules. These methods can increase resource distribution, reduce waiting times, and improve the general efficiency of healthcare service.

Machine learning is a powerful tool that allows healthcare workers to tap the potential of AI in better patient care, optimizing treatment methods, and advancing medical research. By leveraging the vast amounts of healthcare data, machine learning algorithms have the capacity to extract useful insights, causing advancements in personalized medicine, clinical decision support, and community health management.

1.2.3 Natural Language Processing: Enhancing Medical Communication

Natural words Processing (NLP) is an area of AI that works on allowing machines to understand and process human words. It plays a vital part in healthcare by improving medical dialogue, allowing the extraction of useful information from written data, and enabling the development of clever talking bots. Let's study the importance of NLP in healthcare and its uses.

1. Text Processing and Understanding: NLP systems can process and analyze random textual data, such as clinical notes, study books, and patient-generated material. These programs are meant to understand the context, pull important information, and draw meaning from the text. By applying methods such as syntactic parsing, semantic analysis, and named entity recognition, NLP systems can recognize medical ideas, connections, and events from the text, allowing automatic information extraction.

2. Clinical Documentation and Coding: NLP technologies have the ability to simplify and improve the process of clinical documentation and coding. NLP programs can study clinical accounts, take relevant information, and turn it into organized data that can be utilized for billing, quality reports, and research reasons. This technology not only saves time for healthcare workers but also reduces mistakes and improves the accuracy and speed of paperwork processes.

3. Information gathering and books Mining: NLP methods allow the efficient gathering of useful information from big amounts of

medical books and study sources. By utilizing techniques such as information extraction, text classification, and text summary, NLP programs can help academics and healthcare workers in finding the most relevant and up-to-date evidence to guide their treatment decisions. NLP-powered text mining can also help spot new trends, ideas, and knowledge gaps within the medical field.

4. Clinical Decision Support: NLP plays an important role in building clinical decision support systems (CDSS). These systems utilize NLP methods to study patient data, medical books, and clinical standards to provide context-specific suggestions to healthcare workers. NLP-powered CDSS can help in identifying diseases, offering suitable treatment choices, and notifying healthcare workers about possible drug combinations or bad events. By combining NLP with electronic health data, CDSS can offer real-time, targeted, and evidence-based help at the point of care.

5. talking Agents and Chatbots: NLP technologies are crucial in building clever talking agents and chatbots that can connect with patients and healthcare workers. These virtual helpers can understand natural language searches, provide relevant information, plan meetings, answer commonly asked questions, and offer basic medical advice. By leveraging NLP methods, talking bots enhance patient involvement, improve access to healthcare information, and allow personalized exchanges, eventually enabling patients to take an active role in controlling their health.

The application of NLP in healthcare has the potential to change medical communication, improve information access, simplify recording processes, and enhance clinical decision-making. By embracing the power of NLP, healthcare workers can successfully manage the vast amount of written information, draw useful insights, and offer more efficient and patient-centered care.

1.3 Early Applications of AI in Medicine

1.3.1 Diagnostic Algorithms: Pioneering AI in Diagnosing Diseases

One of the early and important uses of AI in medicine is the development of diagnosis algorithms that help healthcare workers in identifying diseases correctly and quickly. These programs harness the power of AI, especially machine learning, to study patient data, understand clinical signs and symptoms, and provide medical insights. Let's study the importance and effect of diagnosis methods in healthcare.

1. Enhanced Accuracy and Efficiency: diagnosis tools driven by AI have the ability to enhance diagnosis accuracy and efficiency. By studying vast amounts of patient data, including medical history, test results, imaging studies, and genetic information, these programs can spot trends, notice minor changes, and make correct diagnoses. The ability of AI systems to handle complex and diverse information allows them to consider a wider range of factors and make links that may be difficult for human doctors alone. This can result in more accurate and quick evaluations, leading to better patient outcomes.

2. Support for Rare and Complex Conditions: AI-based diagnosis systems can provide vital support in identifying rare and complex conditions. These algorithms can examine big datasets, including medical books and expert knowledge, to find specific clinical traits and trends related with these diseases. By leveraging machine learning techniques, diagnostic algorithms can combine diverse data sources and help doctors spot rare diseases, differentiate between related conditions, and guide them towards suitable diagnostic tests and treatments. This can lead to earlier treatments, improved care, and better results for people with rare or difficult diseases.

3. Augmented Decision-Making: Diagnostic algorithms serve as useful tools that assist the decision-making process of healthcare

workers. They can examine a wide range of clinical data and provide doctors with evidence-based suggestions or potential findings. Diagnostic algorithms can act as decision support systems, providing doctors with extra insights, different options, and important clinical standards to consider during the diagnostic process. This joint method between human doctors and AI systems can increase diagnostic trust, reduce diagnostic mistakes, and improve overall patient care.

4. Rapid Triage and Triage Assistance: In situations where time is important, such as emergency rooms or urgent care settings, diagnostic methods can play a vital role in triaging patients. By quickly studying patient data and signs, these programs can rank cases based on the seriousness of the condition, enabling fast treatments and resource allocation. Diagnostic algorithms can also help healthcare workers in emergency decision-making by giving real-time ideas, directing them towards the appropriate diagnostic tests or treatments based on the given signs or data.

5. Scalability and accessible: AI-powered detection systems have the potential to handle the scalability and accessible issues in healthcare. With the growing demand for healthcare services and the lack of skilled medical workers, diagnostic algorithms can help fill the gap by providing scalable diagnostic support. These methods can be applied in various healthcare situations, including rural or impoverished areas, giving access to quality medical help where specialized knowledge may be restricted. This flexibility and ease can improve healthcare equality, especially in areas with low means or healthcare facilities.

The creation and usage of detection algorithms mark a major milestone in the application of AI in medicine. By leveraging the power of AI and machine learning, these systems have the potential to enhance diagnostic accuracy, support rare disease recognition, boost decision-making, allow fast screening, and improve healthcare accessibility. As technology continues to advance, diagnostic programs will likely play an increasingly crucial role in the diagnostic process, paving the way for more personalized and accurate patient care.

1.3.2 Decision Support Systems: Augmenting Medical Decision-Making

In addition to detection algorithms, another early application of AI in medicine is the development of decision support systems (DSS). These systems employ AI technologies, such as machine learning and expert systems, to help healthcare workers in making informed and evidence-based choices. Let's study the importance of decision support tools and their effect on medical decision-making.

1. Integration of Patient Data and Knowledge: Decision support systems combine patient-specific data, medical knowledge, and clinical standards to provide doctors with useful insights and suggestions. These systems can examine electronic health records, test results, medical images, and other relevant data sources to measure patient situations, spot possible risks, and offer suitable treatment choices. By mixing the power of AI algorithms with up-to-date medical knowledge, decision support systems allow doctors to make well-informed decisions tailored to individual patients.

2. Evidence-Based suggestions: Decision support tools focus on evidence-based medicine to create suggestions for healthcare workers. These tools examine a huge amount of medical books, research results, and clinical standards to spot relevant studies and extract relevant information. By utilizing natural language processing and machine learning methods, decision support systems can combine and show the most relevant evidence to doctors, helping them stay informed of the latest research and providing them with a basis for evidence-based decision-making.

3. Clinical process Optimization: Decision support tools can improve the clinical process by simplifying and automating certain chores. These systems can produce alerts for possible drug combinations, notes for prevention screens or vaccines, and ideas for follow-up tests or consults. By handling regular tasks and giving real-time advice, decision support systems help healthcare workers plan and handle patient care more efficiently, reducing the chance of mistakes or lapses in the decision-making process.

4. Personalized Treatment Plans: Decision support systems have the ability to help in building personalized treatment plans for individual patients. By considering patient-specific traits, such as age, gender, medical history, and genetic information, these systems can produce personalized suggestions for tests, medicines, and treatments. Personalized treatment plans drawn from decision support systems aim to improve treatment effectiveness, reduce harmful effects, and consider the unique needs and desires of each patient.

5. Education and Training: Decision support systems can also serve as training tools for healthcare workers, especially for medical students, trainees, and practitioners. These systems can provide engaging learning experiences, case-based models, and virtual patient situations that allow healthcare workers to practice clinical decision-making in a safe setting. By modeling different situations and giving input, decision support systems help to the ongoing education and professional growth of healthcare workers.

The development of decision support systems marks a major improvement in AI uses in healthcare. By combining patient data, medical knowledge, and evidence-based medicine, these systems improve medical decision-making, optimize hospital processes, allow individual treatment plans, and add to healthcare worker education. As AI technologies continue to grow, decision support systems will play an increasingly important part in providing high-quality, patient-centered care and better overall healthcare results.

1.3.3 Robotics in Surgery: Precision and Efficiency in the Operating Room

Robotics has emerged as a transformative tool in the area of surgery, changing the way medical treatments are performed. Robotic surgery systems blend AI, machine learning, and advanced robots to enhance medical accuracy, allow minimally invasive methods, and improve patient results. Let's look into the importance of robots in surgery and its effect on the operating room.

1. Enhanced Precision and agility: Robotic surgical systems provide doctors with enhanced precision and agility, topping the powers of traditional surgical methods. These systems utilize robotic arms controlled by the surgeon, which are equipped with tiny surgical tools and cameras that provide high-definition, enlarged views of the surgery site. The robotic arms offer a greater range of motion, reducing hand shaking and allowing more precise and delicate movements. This accuracy is especially useful for treatments that require complex suturing, dissection, or handling of delicate tissues.

2. Minimally Invasive Techniques: Robot-assisted surgery allows minimally invasive techniques, also known as keyhole surgery, which offer numerous benefits over standard open surgery. Minimally invasive treatments involve making small holes through which robotic tools and a camera are placed. The surgeon controls the robotic arms from a computer, directing the tools with fine movements. This method results in smaller cuts, reduced blood loss, faster healing times, and less post-operative pain for patients. Robotic systems enable minimally invasive surgery by providing the surgeon with enhanced vision, better comfort, and greater accuracy during the process.

3. Surgical Teleoperation and Telesurgery: Robotic surgical systems have the ability to enable surgical teleoperation and telesurgery, allowing doctors to perform treatments afar. In telesurgery, the surgeon controls the robotic arms from a faraway place, giving their knowledge to patients who may not have access to skilled medical care. This potential is especially useful in distant or underdeveloped areas, as it allows patients to receive high-quality medical care without the need for long-distance travel. Additionally, surgical teleoperation allows experienced surgeons to provide direction and mentoring during difficult procedures, promoting surgical skill transfer and cooperation among healthcare workers.

4. surgery Training and Simulation: Robotic surgery systems serve as useful training tools for doctors in training and add to their skill development. These systems offer virtual surgery settings where trainees can practice different techniques in a controlled setting. By utilizing AI and virtual reality technologies, robotic surgical models can recreate real-world surgical scenarios, allowing trainees to gain proficiency in robotic surgical methods, improve their skills, and enhance their decision-making abilities. Robotic models also provide objective performance measurements, feedback, and review, adding to the standards and quality improvement of surgery training.

5. Data record and Analysis: Robotic surgery systems record and analyze vast amounts of medical data, including tool moves, tissue traits, and bodily factors. By applying AI and machine learning methods, this data can be examined to gain insights into surgery techniques, results, and differences in surgical approaches. The study of surgery data adds to the development of evidence-based practices, improvement of surgical processes, and discovery of factors that affect surgical results. This data-driven method has the ability to drive changes in surgery skills, patient safety, and total medical quality.

The merging of robots into the field of surgery has changed the operating room by offering improved accuracy, allowing minimally invasive methods, and easing medical teleoperation. Robotic surgery systems have the ability to change surgical practices, improve patient results, and expand access to high-quality surgical care. As robots technology continues to advance, we can expect further advances that will change the limits of surgical processes and add to the development of surgery science.

AI Revolution in Medicine, The Future of Healthcare

CHAPTER 2: AI AND DIAGNOSTICS

2.1 The Importance of Accurate Diagnostics
2.1.1 The Role of Diagnostics in Healthcare

Accurate diagnostics serve a key role in healthcare by giving important information for illness detection, treatment planning, and monitoring patient progress. Diagnostics cover a variety of tests, treatments, and evaluations that help in diagnosing the underlying cause of a patient's symptoms or health condition. Let's study the crucial function of diagnostics in healthcare.

1. Disease Identification and Early Detection: Diagnostics are vital for detecting illnesses and health issues properly. They help healthcare practitioners to discern between distinct illnesses, estimate their severity, and make educated judgments about treatment alternatives. By identifying illnesses in their early stages, diagnostics permit timely interventions, enhancing the likelihood of effective treatment and improved patient outcomes.

2. therapy Planning and Monitoring: Accurate diagnoses give crucial information that informs therapy planning. Diagnostic tests assist healthcare providers determine the features of an illness, its stage, and its possible response to various therapies. This information enables for the formulation of individualized treatment programs suited to specific patients, enhancing therapeutic efficacy and avoiding needless procedures. Moreover, diagnostics enable constant monitoring of a patient's reaction to therapy, allowing for tweaks or revisions as required.

3. Prognostic Assessment and Risk Stratification: Diagnostics help in prognostic assessment by forecasting the probable course and outcome of a disease. By evaluating numerous indicators, such as biomarkers, genetic variables, or imaging data, diagnostics may

give insights into the prognosis of a certain ailment. This information supports healthcare workers in identifying the proper degree of treatment, forecasting probable consequences, and stratifying patients depending on their risk profile.

4. Screening and Preventive Medicine: Diagnostics play a significant role in screening for illnesses and identifying persons at risk before symptoms arise. Screening procedures, such as mammograms, colonoscopies, or blood tests, may discover illnesses in asymptomatic persons, allowing early interventions and better treatment results. Moreover, diagnostics contribute in preventive medicine by detecting risk factors and allowing actions to lower the possibility of disease development, such as lifestyle adjustments or vaccines.

5. Public Health Surveillance and Outbreak control: Diagnostics are vital for public health surveillance, especially in the identification and control of infectious illnesses. Diagnostic tests allow the detection and tracking of disease outbreaks, letting public health authorities execute early actions to restrict the spread of illnesses. Additionally, diagnostics offer vital data for monitoring illness patterns, assessing the efficiency of public health initiatives, and shaping public health policy.

Accurate diagnoses are the cornerstone of successful healthcare delivery. They allow rapid illness detection, guide treatment choices, assist preventative initiatives, and contribute to public health monitoring. The ongoing improvement of diagnostic technology, along with the incorporation of AI, offers significant promise to further boost diagnostic accuracy, accessibility, and precision medicine techniques. By concentrating on enhancing diagnostics, we can drastically improve patient outcomes and alter healthcare delivery.

2.1.2 Challenges in Diagnostic Accuracy

While precise diagnoses are critical for successful healthcare, various difficulties exist that might impair diagnostic accuracy. These obstacles originate from several sources, including the complexity of illnesses, limits of diagnostic technology, and intrinsic heterogeneity in human interpretation. Understanding these issues is vital for devising methods to overcome them and enhance diagnostic accuracy. Let's review some of the major issues in diagnostic accuracy:

1. Variability and Subjectivity: Interpretation of diagnostic tests sometimes includes subjective judgment, resulting to variability across healthcare providers. Different specialists may interpret test findings differently, possibly leading to disparities in diagnosis. This diversity may be impacted by variables such as experience, knowledge, biases, and intrinsic subjectivity in the interpretation process. Addressing this difficulty involves initiatives to standardize diagnostic criteria, enhance inter-observer agreement, and employ technology to offer objective and consistent diagnostic evaluations.

2. Limited Access to Diagnostics: In many regions of the globe, access to diagnostic facilities and technology is limited. This may lead to delays in diagnosis and treatment start, especially in resource-constrained situations. Limited access to diagnostics disproportionately impacts underprivileged communities and may result in discrepancies in healthcare outcomes. Efforts to expand access to diagnostics, especially in low-resource areas, are vital to promote fair healthcare delivery and boost diagnostic accuracy on a worldwide scale.

3. Complexity and Multifactorial Nature of illnesses: Many illnesses are complex and multifactorial in nature, creating diagnostic problems. Some illnesses have overlapping symptoms or cryptic presentations, making correct diagnosis challenging. Additionally, illnesses may include several interacting elements, such as genetic, environmental, and lifestyle factors, further

complicating the diagnosis procedure. Advanced diagnostic procedures, including the incorporation of AI algorithms, may assist navigate the complexity of illnesses and enhance diagnosis accuracy by examining a broad variety of factors concurrently.

4. Inadequate Training and Continuing Education: Diagnostic accuracy significantly depends on the abilities and knowledge of healthcare professionals participating in the diagnostic process. However, insufficient training or restricted access to continuing education opportunities might hamper diagnosis accuracy. Keeping healthcare workers up-to-date with the newest breakthroughs, standards, and best practices in diagnostics is vital to increase accuracy. Investments in training programs, knowledge-sharing platforms, and continuing professional growth may help to increasing diagnostic abilities and lowering diagnostic mistakes.

5. Technical limits and False Results: Diagnostic tests and technologies may have inherent limits that might impair accuracy. False-positive or false-negative findings may arise owing to several variables, including technical mistakes, equipment constraints, specimen quality, and assay sensitivity. These incorrect findings might lead to misdiagnosis or delayed diagnosis, compromising patient outcomes. It is necessary to invest in comprehensive quality assurance processes, standardized methods, and thorough validation of diagnostic technologies to limit the incidence of false findings and boost diagnostic accuracy.

Addressing these difficulties needs a multi-faceted strategy that combines technology breakthroughs, quality improvement initiatives, education and training programs, and efforts to enhance healthcare infrastructure and accessibility. By addressing these hurdles, healthcare systems may boost diagnostic accuracy, minimize diagnostic mistakes, and ultimately improve patient outcomes. The incorporation of AI technology in diagnostics offers promise in tackling some of these difficulties by delivering objective and standardized diagnostic evaluations, aiding in complicated decision-making, and complementing the capacities of healthcare personnel.

2.2 AI in Diagnostic Imaging

2.2.1 Evolution of AI in Diagnostic Imaging

Diagnostic imaging serves a key role in healthcare, allowing the viewing and characterization of anatomical structures, physiological processes, and disease states. The incorporation of artificial intelligence (AI) in diagnostic imaging has revolutionised the discipline, altering the way medical pictures are interpreted, processed, and used for diagnosis and treatment planning. Let's investigate the progress of AI in diagnostic imaging and its influence on healthcare.

1. Early Developments: The early usage of AI in diagnostic imaging may be traced back to the 1960s when researchers started investigating the possibility of computer-based image processing. Initially, AI algorithms concentrated on fundamental tasks such as picture improvement and feature extraction. Over time, developments in processing power and machine learning methods enabled for increasingly complex applications in diagnostic imaging.

2. Machine Learning and Deep Learning: The introduction of machine learning and deep learning methods has considerably increased the capabilities of AI in diagnostic imaging. Machine learning methods allow computers to learn from big datasets and make predictions or judgments without explicit programming. Deep learning, a form of machine learning, employs artificial neural networks with numerous layers to extract complicated patterns and characteristics from medical pictures.

3. picture Classification and Segmentation: AI algorithms have been effectively used to picture classification and segmentation

tasks in diagnostic imaging. For example, convolutional neural networks (CNNs) have exhibited amazing ability in categorizing and recognizing particular anatomical structures or abnormalities inside medical pictures. AI-based segmentation algorithms can correctly identify areas of interest, assisting in tumor identification, organ volume assessments, and treatment planning.

4. Computer-Aided identification and Diagnosis: AI algorithms have showed promise in aiding radiologists and doctors with the identification and diagnosis of illnesses in medical pictures. Computer-aided detection (CAD) systems utilize AI algorithms to scan pictures and indicate probable anomalies, supporting radiologists in identifying regions that need additional attention. CAD systems have proved especially efficient in identifying early-stage malignancies, such as breast and lung tumors, boosting diagnosis accuracy and assisting in early management.

5. Quantitative Image Analysis and Radiomics: AI has allowed the extraction of quantitative information from medical pictures, leading to the area of radiomics. Radiomics includes the high-throughput extraction and analysis of a large number of quantitative information from pictures. AI algorithms may examine these radiomic data to give further insights into illness characteristics, therapy response, and prognosis evaluation. Radiomics has the potential to help personalized medicine methods by discovering imaging biomarkers and aiding in treatment decision-making.

6. Integration with healthcare Workflows: AI technologies are rapidly being incorporated into healthcare workflows to expedite and improve diagnostic imaging operations. AI systems can triage and prioritize pictures, pre-process images for radiologists, and give automated measurements or comments. The integration of AI in Picture Archiving and Communication Systems (PACS) and Radiology Information Systems (RIS) enables for seamless

application of AI algorithms inside the radiology workflow, enhancing efficiency and diagnostic accuracy.

The growth of AI in diagnostic imaging has revolutionised the sector, enabling healthcare practitioners with improved tools and procedures to increase diagnostic accuracy, efficiency, and patient care. The incorporation of AI algorithms in diagnostic imaging has enormous potential for early illness identification, accurate diagnosis, and individualized therapy planning. Continued breakthroughs in AI technology, data availability, and partnerships between AI specialists and doctors are vital to unlocking the full potential of AI in diagnostic imaging and its incorporation into everyday clinical practice.

2.2.2 Applications of AI in Radiology

2.2.2.1 Automated Image Analysis

Automated image analysis is one of the primary uses of artificial intelligence (AI) in radiology. By employing AI algorithms, radiologists may automate the study and interpretation of medical images, enhancing productivity and diagnostic accuracy. Let's discuss the numerous features and advantages of automated image analysis in radiology.

1. Image Preprocessing and Enhancement: AI algorithms may be used to preprocess and enhance medical images, enhancing their quality and aiding better interpretation. For example, AI approaches may decrease noise, repair artifacts, and boost picture contrast, enhancing the appearance of anatomical features and anomalies. This preprocessing procedure benefits radiologists in acquiring better and more informative pictures, leading to more accurate diagnoses.

2. Lesion Detection and Localization: AI systems excel in identifying and localizing anomalies or lesions within medical

pictures. By examining patterns, textures, and forms, AI systems can automatically find and flag possible areas of concern. This allows radiologists to concentrate their attention on locations that need additional investigation, lowering the chance of missing crucial discoveries and boosting overall diagnostic accuracy.

3. Quantitative Analysis and Measurement: Automated image analysis enables for exact and quantitative measures of anatomical features or lesions inside medical pictures. AI algorithms can properly determine metrics such as size, volume, density, or blood flow characteristics. This quantitative analysis supports in disease staging, treatment planning, and tracking patient response to therapy. It also permits objective comparisons of pictures across time, permitting longitudinal examinations.

4. Computer-Aided Diagnosis (CAD): Computer-aided diagnosis (CAD) systems apply AI algorithms to give diagnostic support to radiologists. CAD systems analyze medical pictures and deliver computer-generated conclusions or recommendations based on patterns and anomalies discovered in the images. These technologies act as a "second pair of eyes," supporting radiologists in their decision-making process and perhaps minimizing diagnostic mistakes. CAD systems have been especially helpful in identifying early-stage malignancies, such as breast cancer and lung cancer.

5. Risk Stratification and Prognostic Assessment: AI algorithms can evaluate medical pictures and extract characteristics that are indicative of disease aggressiveness, risk stratification, and prognosis. By measuring certain features of tumors or lesions, AI algorithms may aid in assessing the probability of disease progression or the possible response to alternative treatment choices. This information promotes customized medicine techniques, allowing for customised treatment strategies based on unique patient characteristics.

6. Workflow Optimization: Automated image analysis assists to workflow optimization in radiology departments. AI systems can triage and prioritize pictures depending on urgency or complexity, allowing radiologists to concentrate on urgent cases. They may also aid in creating preliminary reports, annotating photos, and organising data for easier incorporation into electronic health records (EHRs). By automating time-consuming procedures, AI boosts productivity and enables radiologists to spend more time on difficult interpretations and patient care.

Automated image analysis driven by AI has the potential to transform radiology by boosting efficiency, accuracy, and patient outcomes. By employing AI algorithms to help with image interpretation, radiologists may make more informed diagnoses, minimize mistakes, and increase the overall quality of healthcare delivery in radiology. Continued breakthroughs in AI technology, along with strong validation and integration into clinical processes, will further boost the area of automated image processing in radiology.

2.2.2.2 Computer-Aided Detection and Diagnosis

Computer-aided detection and diagnosis (CAD) is a potent use of artificial intelligence (AI) in radiography. CAD systems apply AI algorithms to evaluate medical pictures and give automated support to radiologists in finding and diagnosing problems. Let's study the relevance and advantages of computer-aided detection and diagnosis in radiology.

1. Abnormality Detection: CAD systems are meant to identify and highlight probable abnormalities or areas of interest within medical pictures. By examining patterns, shapes, and textures, AI systems may find subtle or early indicators of illnesses that may be tough to notice visually. This involves recognizing lesions, masses,

microcalcifications, or other aberrant formations that may signal the presence of illnesses such as cancer or cardiovascular disorders.

2. Improved Sensitivity and Accuracy: Computer-aided detection augments radiologists' skills by enhancing sensitivity and accuracy in identifying abnormalities. CAD systems may uncover results that may have been ignored or neglected, minimizing the chance of false negatives. The combination of radiologists' experience with CAD's capacity to analyze enormous volumes of data boosts overall diagnosis accuracy and enables full examination of medical pictures.

3. Second Opinion and Decision assistance: CAD systems operate as a helpful second opinion tool, giving extra insights and decision assistance to radiologists. By evaluating the image data, CAD algorithms provide computer-generated discoveries or recommendations that might help in radiologists' interpretation and decision-making process. This partnership between AI and human professionals boosts diagnostic confidence and helps eliminate diagnostic mistakes.

4. Workflow Efficiency: CAD systems assist to workflow efficiency in radiology departments. They aid in prioritizing cases, enabling radiologists to concentrate their attention on crucial or difficult discoveries. CAD algorithms can automatically identify questionable spots in photos, accelerating the identification and review process. By improving workflow and lowering time spent on common procedures, CAD systems boost radiologists' productivity and allow prompt patient treatment.

5. Education and Training: Computer-aided detection and diagnostic systems serve as important training tools for radiology trainees and residents. CAD methods may be used to produce instructional datasets and recreate real-world settings, allowing trainees to practice interpretation abilities and learn from computer-

generated feedback. This boosts their learning experience, competency, and diagnostic capacities, eventually benefitting patient care.

6. Research and Data study: CAD systems create huge volumes of data from the study of medical pictures. This data may be exploited for research objectives, such as studying illness patterns, discovering imaging biomarkers, and understanding therapy responses. The integration of CAD systems with electronic health records (EHRs) improves data mining and population-based investigations, leading to breakthroughs in radiology and clinical research.

Computer-aided detection and diagnosis tools have revolutionised radiology practice by enhancing accuracy, efficiency, and diagnostic confidence. The synergy between AI algorithms and radiologists' experience boosts the capabilities of both, leading to improved patient outcomes. Continued improvements in AI technology and integration of CAD systems into clinical workflows offer enormous potential for significantly increasing diagnostic capabilities and changing radiology practice.

2.2.2.3 Image Segmentation and Classification

Image segmentation and classification are crucial components of computer-aided detection and diagnosis in radiology. Through the use of artificial intelligence (AI) algorithms, these approaches play a significant role in interpreting medical pictures and supporting radiologists in identifying and describing particular areas or structures of interest. Let's look into the relevance and advantages of picture segmentation and classification in radiology.

1. Image Segmentation: Image segmentation entails separating medical pictures into relevant sections or segments based on their visual features. AI algorithms examine the visual data and find borders or outlines that distinguish distinct anatomical structures or

lesions. This procedure assists in isolating certain portions for additional research, measurement, or visualization. Accurate picture segmentation helps radiologists to concentrate on specified regions and aids accurate localization and characterisation of abnormalities.

2. Organ and Tumor Delineation: Image segmentation is important in defining organs, tissues, or malignancies within medical imaging. AI algorithms may automatically delineate and segment numerous anatomical structures, such as the heart, liver, lungs, or brain. This allows quantitative analysis, volume measurements, and treatment planning. In oncology, tumor segmentation permits accurate identification of tumor borders, assisting in staging, assessment of treatment response, and radiation therapy planning.

3. Lesion characterisation: Image segmentation mixed with classification algorithms allows for the characterisation of lesions or abnormalities observed in medical imaging. AI algorithms examine the segmented areas and extract significant information such as shape, texture, or intensity patterns. These traits may be used to identify lesions as benign or malignant, assess the probability of various diseases, or predict prognosis. Accurate lesion characterisation improves diagnostic decision-making and therapy planning.

4. Quantitative measures: Segmentation methods aid to getting quantitative measures from medical pictures. By properly identifying structures or areas of interest, AI algorithms may determine factors such as size, volume, density, or blood flow characteristics. Quantitative measures give objective and standardized information, allowing accurate monitoring of illness development, response to medication, or treatment planning. These metrics aid in individualized patient care and promote evidence-based medicine.

AI Revolution in Medicine, The Future of Healthcare

5. Automated Classification and Risk Stratification: AI systems can categorize medical pictures into multiple categories or risk groups based on visual criteria. For example, in mammography, AI may identify breast tumors as benign, worrisome, or malignant, supporting radiologists in prioritizing patients and advising proper care choices. Automated categorization and risk stratification can expedite workflow, assure early intervention for high-risk situations, and enhance patient outcomes.

6. Integration with Treatment Planning: Accurate picture segmentation and classification algorithms play a significant role in treatment planning and radiation therapy. By accurately identifying organs at risk and tumor sizes, AI systems aid in optimizing radiation dosage delivery while limiting harm to adjacent healthy tissues. Integration of image segmentation with treatment planning systems increases the accuracy and effectiveness of radiation therapy, increasing patient outcomes and lowering side effects.

Image segmentation and classification approaches driven by AI have transformed radiology by providing exact analysis, characterisation, and quantification of medical images. These strategies boost diagnostic accuracy, permit tailored treatment planning, and expedite workflow. Continued improvements in AI algorithms, integration with clinical processes, and joint research initiatives offer great promise for further boosting picture segmentation and classification in radiology practice.

2.3 AI in Pathology and Laboratory Medicine
2.3.1 Transforming Pathology with AI

Pathology plays a significant role in identifying illnesses and directing therapy options. The integration of artificial intelligence (AI) with pathology and laboratory medicine has transformed the discipline, revolutionizing the way tissue samples and laboratory tests are evaluated and interpreted. Let's study how AI is transforming pathology and its influence on patient care.

1. Automated Image Analysis: AI systems have made considerable gains in automating image analysis in pathology. Digital pathology allows for the digitalization of glass slides, which may be evaluated using AI algorithms. By employing machine learning and deep learning approaches, AI may aid pathologists in tasks such as tissue detection, cell categorization, and quantification of cellular characteristics. This technology saves time, boosts accuracy, and allows pathologists to concentrate on more complicated cases.

2. Faster and More Accurate Diagnoses: AI in pathology has the ability to speed the diagnosis process and increase accuracy. AI systems can swiftly examine vast datasets and uncover patterns or abnormalities that may not be immediately observable by human pathologists. This may lead to speedier and more accurate diagnoses, especially in circumstances when time is crucial, such as cancer detection or the identification of infectious organisms.

3. Precision Medicine and Personalized therapy: AI in pathology helps the progress of precision medicine by offering insights into disease subtypes, prognostic variables, and therapy responses. AI algorithms can examine histopathology pictures and genomic data to find particular biomarkers or genetic abnormalities linked with distinct illness outcomes. This information assists in personalizing treatment programs to specific patients, enhancing therapeutic effectiveness, and avoiding possible adverse effects.

4. Predictive and Prognostic Models: AI algorithms may construct predictive and prognostic models by merging pathology data with clinical data and patient outcomes. By examining large-scale information, AI may find correlations, patterns, and prediction indicators that may not be obvious using standard analytic

approaches. These models may aid in forecasting disease progression, therapy responses, and patient survival, allowing better informed decision-making and improved patient care.

5. Quality Assurance and Standardization: AI algorithms may aid to quality assurance and standardization in pathology. By offering automatic checks and comparisons, AI can assure consistency in diagnosis, adherence to standards, and compliance with best practices. This helps decrease inter-observer variability and increase overall diagnostic accuracy. AI may also aid in auditing and monitoring laboratory procedures, boosting quality control methods.

6. cooperation and Telepathology: AI technology promotes cooperation among pathologists, regardless of geographical boundaries. Telepathology systems linked with AI algorithms enable for remote consultation, second views, and information exchange. Pathologists may employ AI-powered technologies to collaborate on tough cases, share ideas, and obtain expert viewpoints, therefore boosting diagnosis accuracy and lowering diagnostic mistakes.

The integration of AI in pathology and laboratory medicine has significant potential for increasing diagnosis accuracy, efficiency, and patient outcomes. By merging the experience of pathologists with the capabilities of AI algorithms, the discipline is evolving towards more customized and accurate therapy. Continued study, validation, and deployment of AI technologies in pathology will help to its continuous transformation and the delivery of high-quality healthcare.

2.3.2 Applications of AI in Pathology

2.3.2.1 Digital Pathology and Whole Slide Imaging

Digital pathology and entire slide imaging are among the primary uses of artificial intelligence (AI) in pathology. These technologies have transformed the way pathology samples are evaluated, stored, and communicated. Let's investigate the relevance and advantages of digital pathology and whole slide imaging in the perspective of AI.

1. Digitalization of Pathology Samples: Digital pathology is the conversion of glass slides carrying tissue samples into digital pictures. High-resolution scanners capture the whole slide, retaining all cellular and architectural features. This digitization removes the requirement for physical storage and enables for quick access to pathology samples from anywhere in the globe. It also permits integration with AI algorithms for automated analysis.

2. Whole Slide Imaging (WSI): Whole slide imaging (WSI) refers to the method of digitizing an entire glass slide at high magnification. WSI scanners capture the whole slide as a single digital picture, retaining all cellular and structural information. This allows pathologists to traverse and study the slide digitally, zooming in and out to analyze particular areas of interest. WSI is the cornerstone of digital pathology and AI-assisted analysis.

3. AI-Assisted Analysis: AI algorithms may be used to digital pathology pictures to aid pathologists in different activities. By utilizing machine learning and deep learning approaches, AI systems may evaluate digital pathology pictures and give automated help in areas such as tissue identification, cell segmentation, feature extraction, and classification. This partnership between AI and pathologists promotes diagnosis accuracy, speed, and consistency.

4. Tumor Detection and Quantification: AI systems can automatically discover and quantify tumors inside digital pathology pictures. By assessing cellular properties, tissue patterns, and morphological traits, AI can identify areas of interest and evaluate

tumor size, volume, and density. This information assists in tumor staging, prognosis evaluation, and therapy planning. AI-assisted tumor identification and quantification save time and minimize inter-observer variability.

5. Image Analysis and Pattern Recognition: AI systems excel in analyzing massive volumes of digital pathology pictures and identifying patterns or abnormalities that may be tough for human observers to discover. By learning from enormous datasets, AI systems may spot minor signs, such as nuclear abnormalities, mitotic figures, or architectural modifications that may signal particular illnesses or prognostic variables. This research boosts diagnosis accuracy and gives useful insights for treatment options.

6. Education and Training: Digital pathology and AI-assisted analysis are useful instructional materials for pathology trainees and residents. Digital slides may be shared and viewed remotely, allowing trainees to study and learn from a broad variety of situations. AI algorithms may aid in building training datasets, modeling diagnostic situations, and offering feedback on trainees' interpretations. This increases learning, proficiency, and uniformity in pathology practice.

7. Remote Consultations and Collaborations: Digital pathology and AI-assisted analysis offer remote consultations and collaborations among pathologists. Pathologists may exchange digital slides with colleagues for second views, expert consultations, or multidisciplinary conversations. AI algorithms may help in automatic annotations, identifying locations of interest or probable irregularities, allowing efficient and productive remote collaborations.

Digital pathology and whole slide imaging, paired with AI-assisted analysis, have changed the science of pathology. These technologies boost diagnostic accuracy, permit remote consultations, simplify education and training, and contribute to customized medical methods. Continued breakthroughs in digital pathology and AI algorithms will further drive innovation, enhancing patient care and results in pathology.

2.3.2.2 AI-Assisted Diagnosis in Pathology

AI-assisted diagnosis is a pioneering use of artificial intelligence (AI) in pathology that revolutionizes the area of illness diagnosis. By employing modern algorithms and machine learning approaches, AI can give substantial support to pathologists in analyzing pathology samples and providing correct diagnosis. Let's look into the relevance and advantages of AI-assisted diagnosis in pathology.

1. Increased Diagnostic Accuracy: AI systems can evaluate digital pathology pictures and aid pathologists in making more accurate diagnoses. By learning from enormous datasets and spotting minor trends or irregularities, AI may highlight possible areas of concern or offer differential diagnosis. This partnership between AI and pathologists boosts diagnosis accuracy, eliminates diagnostic mistakes, and assures prompt and appropriate patient care.

2. quicker Turnaround Time: AI-assisted diagnostics speeds the diagnostic procedure, leading to quicker turnaround times. Pathologists may employ AI algorithms to prioritize cases, uncover crucial results, and triage samples that demand quick care. By automating key processes and offering early assessments, AI minimizes the stress on pathologists and improves workflow, allowing quick diagnosis and treatment commencement.

3. Identification of Rare or Uncommon illnesses: AI algorithms may help in the identification and diagnosis of rare or uncommon illnesses. By evaluating digital pathology photos and comparing them with huge databases, AI can discover patterns and traits suggestive of certain illnesses, even ones that pathologists may see

seldom. This tool helps pathologists traverse difficult situations and assures correct diagnosis, leading to suitable treatment methods.

4. Decision Support and Treatment Guidance: AI-assisted diagnosis gives essential decision support to pathologists in choosing effective treatment methods. By evaluating pathology samples, AI systems may offer individualized therapy choices, predict treatment responses, or uncover prospective therapeutic targets. This information aids patient care by providing targeted therapy and increasing treatment results.

5. Quality Assurance and Standardization: AI algorithms aid to quality assurance and standardization in pathology practice. By automating specific activities and giving consistent judgments, AI helps decrease inter-observer variability and assures adherence to established rules and best practices. This boosts diagnostic accuracy, enhances quality control, and adds to the general dependability of pathology diagnosis.

6. Integration with Electronic Health Records (EHRs): AI-assisted diagnosis may be effortlessly connected with electronic health records (EHRs), allowing efficient data management and supporting data-driven healthcare. AI algorithms may extract essential information from pathology reports and combine it with other clinical data, offering a full perspective of the patient's health state. This integration promotes evidence-based care, research, and community health management.

7. Continuous Learning and Improvement: AI-assisted diagnostic systems may continually learn and improve over time. By assessing comments from pathologists, comparing diagnoses with gold standards, and adding fresh data, AI systems may enhance their diagnostic skills. This repeated learning process boosts the accuracy and reliability of AI-assisted diagnoses, assuring continual progress in patient care.

AI Revolution in Medicine, The Future of Healthcare

AI-assisted diagnosis in pathology offers enormous promise to boost diagnostic accuracy, speed up the diagnostic process, and improve patient outcomes. The synergy between AI algorithms and pathologists' knowledge leads to a dynamic partnership that benefits both pathologists and patients. Continued study, validation, and integration of AI technologies into pathology practice will pave the road for greater improvements in diagnostics and tailored treatment.

2.3.2.3 Predictive Analytics and Risk Assessment

In the realm of pathology, the integration of artificial intelligence (AI) has brought up new opportunities for predictive analytics and risk assessment. AI algorithms may examine pathology data, patient information, and other relevant data sources to produce insights and predictions about disease development, treatment responses, and patient outcomes. Let's investigate the relevance and uses of predictive analytics and risk assessment in pathology.

1. Disease development and Prognosis: AI algorithms may employ large-scale pathology information to uncover patterns and parameters related with disease development. By assessing histopathological pictures, genetic data, and clinical characteristics, AI can predict the risk of disease development, metastasis, or recurrence. This information assists in predicting prognosis, adjusting therapy strategies, and monitoring patients for early intervention.

2. therapy Response Prediction: AI systems can analyze pathology samples and patient data to predict therapy responses. By assessing cellular features, molecular markers, and clinical data, AI can forecast the probability of good treatment results or indicate possible resistance to certain medications. This helps doctors to make educated judgments about therapy selection, dose modifications, and alternative therapeutic alternatives.

3. Risk Stratification and Early Detection: AI algorithms may assist in risk stratification and early detection of illnesses. By examining pathology data, patient demographics, genetic markers, and environmental variables, AI may identify people at increased risk of getting specific illnesses. This permits focused screening programs, proactive treatments, and early discovery of preclinical phases, when treatment choices are more successful.

4. individualized Risk Assessment: AI systems may conduct individualized risk assessment by merging individual patient data with population-based information. By analyzing genetic predispositions, lifestyle variables, and medical history, AI may predict an individual's risk of getting certain illnesses or suffering unfavorable health occurrences. This information supports individualized preventative actions, early interventions, and customised healthcare planning.

5. Population Health Management: AI algorithms may contribute to population health management by evaluating pathology data and discovering trends at a population level. By recognizing illness patterns, risk factors, and regional differences, AI can lead public health efforts, resource allocation, and preventative methods. This helps early intervention, health promotion, and the optimization of healthcare resources.

6. Clinical Decision assistance: AI algorithms may give decision assistance to pathologists and clinicians by delivering appropriate risk assessment and predictive analytics data. This facilitates in the interpretation of pathology data, selection of relevant diagnostic tests, and creation of treatment programs. AI-powered decision support systems increase clinical decision-making, leading to better patient care and results.

7. Research and medication Development: AI-assisted predictive analytics may enhance research efforts and medication development

in pathology. By analyzing massive datasets and finding molecular targets, AI algorithms may help in the discovery and development of novel medicines. Additionally, AI may aid in patient classification for clinical trials, allowing the identification of potential candidates based on projected treatment responses.

The combination of predictive analytics and risk assessment with AI in pathology has the potential to improve patient treatment, disease management, and public health measures. By using the power of AI algorithms, pathologists and healthcare professionals may make more educated judgments, administer individualized therapies, and enhance patient outcomes. Continued breakthroughs and cooperation in AI research and implementation will further strengthen the predictive powers of pathology, contributing to precision medicine and public health gains.

2.4 AI in Diagnostic Decision Support

2.4.1 Enhancing Diagnostic Accuracy with AI

diagnosis accuracy is a critical aspect of healthcare, and the combination of artificial intelligence (AI) has shown great promise in improving diagnosis accuracy across various medical fields. AI programs can examine vast amounts of patient data, medical pictures, test results, and clinical information to provide useful decision support to healthcare workers. Let's study how AI improves diagnosis accuracy in healthcare.

1. Pattern Recognition and Data Analysis: AI systems shine at spotting complicated patterns and studying big datasets. By working on extensive sets of medical pictures, AI can spot minor flaws or traits that may be difficult for human viewers to discover. This pattern recognition capability improves testing accuracy by providing additional insights and helping healthcare workers in making more informed decisions.

2. improved Interpretation and Second views: AI systems can provide improved interpretation and second views to healthcare workers. By studying patient data and medical pictures, AI can create suggestions or show areas of concern, easing the testing process. This partnership between AI and healthcare workers improves accuracy by lowering diagnosis mistakes and offering additional views.

3. Differential Diagnosis Support: AI systems can help in getting differential diagnoses by studying patient data and showing possible diagnoses based on the available information. By considering various clinical factors, symptoms, and test results, AI can offer a range of possible conditions, helping healthcare workers in cutting down their diagnostic considerations. This help lowers diagnostic confusion and improves precision.

4. Risk Assessment and Prognostic Prediction: AI systems can assess patient data and medical records to provide risk assessment and prognostic predictions. By studying various factors such as patient data, medical history, and test results, AI can predict the chance of disease development, treatment reaction, or bad effects. This knowledge helps healthcare workers in making informed choices and giving personalized care.

5. Integration of Multimodal Data: AI systems can combine and evaluate data from multiple sources, such as medical pictures, genetic data, electronic health records, and smart devices. By mixing these diverse data types, AI can provide a complete view of the patient's health state and help in correct evaluation. This combination improves medical accuracy by considering a complete picture of the patient's state.

6. Real-Time Decision Support: AI systems can provide real-time decision support to healthcare workers at the point of care. By studying patient data in real-time, AI can help in the analysis of

results, suggest suitable diagnosis tests, and provide treatment suggestions based on the latest evidence. This real-time decision support improves testing accuracy by ensuring up-to-date and evidence-based care.

7. Continuous Learning and Improvement: AI systems can continuously learn and improve their diagnostic skills through feedback and exposure to new data. By adding comments from healthcare experts and comparing findings with gold standards, AI algorithms can improve their algorithms and enhance diagnostic accuracy over time. This iterative learning method ensures ongoing improvement and adaptation to changing medical knowledge.

The inclusion of AI in medical decision support holds great potential for improving diagnosis accuracy in healthcare. By leveraging AI systems' ability to study big datasets, spot patterns, and provide decision support, healthcare workers can benefit from better accuracy, reduced medical mistakes, and enhanced patient results. Continued study, confirmation, and cooperation between AI and healthcare workers will drive further breakthroughs in medical decision support, changing the field of medicine.

2.4.2 Clinical Decision Support Systems (CDSS)

2.4.2.1 Integration of AI Algorithms in CDSS

Clinical Decision Support Systems (CDSS) play a vital part in healthcare by helping healthcare workers in making well-informed decisions at the point of care. With the merging of artificial intelligence (AI) tools, CDSS has seen significant improvements in increasing diagnosis accuracy and better patient results. Let's study the inclusion of AI systems in CDSS and its effect on healthcare decision-making.

1. Data Integration and Analysis: AI methods built into CDSS can analyze and combine diverse patient data, including medical

records, test reports, medical images, and genome information. By leveraging AI's ability to process and analyze big datasets, CDSS can provide thorough and up-to-date information to healthcare workers, supporting medical decision-making.

2. Risk Stratification and predicted Analytics: AI systems in CDSS can review patient data to stratify risk and provide predicted analytics. By considering various clinical factors, patient history, and population-based data, AI systems can predict the chance of disease development, bad events, or treatment reaction. This allows healthcare workers to make educated decisions about suitable treatments and individual care plans.

3. Decision Support for Diagnosis and Treatment: AI systems built into CDSS can provide decision support for diagnosis and treatment plans. By studying patient data and medical knowledge libraries, CDSS can give suggestions on alternative diagnoses, suitable diagnosis tests, and treatment choices. This support helps healthcare workers in making accurate and quick decisions, leading to better patient results.

4. Real-Time Alerts and messages: AI systems in CDSS can watch patient data in real-time and provide alerts or messages to healthcare workers for important events, such as odd test results or drug combinations. By leveraging AI's ability to process data quickly, CDSS can ensure timely responses and prevent medical mistakes, thereby improving patient safety.

5. proof-Based Medicine: AI systems built into CDSS can constantly learn from new medical study and proof. By improving their knowledge base, CDSS can provide healthcare workers with the most current evidence-based suggestions and standards. This guarantees that professional choices match with the latest medical knowledge, leading to better diagnosis accuracy and best patient care.

AI Revolution in Medicine, The Future of Healthcare

6. Workflow Integration: AI-integrated CDSS can be easily merged into the clinical workflow, giving decision help at the point of care. By merging with electronic health record systems and other healthcare tools, CDSS can access important patient information and provide context-specific suggestions. This combination simplifies professional decision-making and boosts speed in healthcare service.

7. Learning and Adaptation: AI systems in CDSS can constantly learn and change based on input and results data. By studying the results of choices made with CDSS help, AI systems can hone their suggestions and improve their accuracy over time. This iterative learning process guarantees that CDSS stays up-to-date and matched with the changing medical world.

The inclusion of AI methods in CDSS holds great promise for improving diagnosis accuracy, treatment plans, and patient results. By leveraging AI's skills in data analysis, pattern recognition, and real-time decision support, CDSS helps healthcare workers in making well-informed choices. Continued study, evaluation, and teamwork between AI experts and healthcare workers will drive further breakthroughs in AI-integrated CDSS, eventually helping patients and changing healthcare service.

2.4.2.2 Case-Based Reasoning and Clinical Guidelines

In the world of clinical decision support systems (CDSS), the merging of artificial intelligence (AI) programs has brought new methods such as case-based thinking and the usage of clinical standards. These methods improve testing accuracy and help healthcare workers in making informed choices. Let's study the merging of AI systems in CDSS through case-based logic and clinical standards.

1. Case-Based Reasoning: Case-based reasoning (CBR) includes utilizing AI systems to examine and compare current patient cases

with previously met cases saved in a knowledge base. This method helps CDSS to provide context-specific suggestions based on similar cases, enabling diagnosis accuracy and treatment plans. By considering patterns in symptoms, medical background, and test results, CBR can help healthcare workers in finding possible illnesses and suitable treatment strategies.

2. Knowledge Base and Case Retrieval: CDSS utilizing case-based reasoning rests on a thorough knowledge base that stores a vast collection of anonymous patient cases. These cases cover diverse medical scenarios, allowing the AI algorithms to find relevant cases based on the input data and compare them with the present patient's traits. The recovered cases provide useful insights, helping healthcare workers in the diagnosis process and treatment choices.

3. resemblance Assessment and Adaptation: Once relevant cases are retrieved, AI systems measure the resemblance between the current patient's case and the recovered cases. Similarity is judged based on various factors such as symptoms, test results, medical background, and personal information. The AI systems change and apply the information from the related cases to the current patient's situation, giving possible diagnoses, treatment choices, and future predictions.

4. Learning from Outcomes: Case-based thinking in CDSS includes a feedback system that allows learning from outcomes. By studying the results of previously suggested actions, the AI systems can change and improve their suggestions over time. This repeated learning process improves the accuracy and usefulness of the CDSS, leading to constant improvement and optimization of diagnosis and treatment ideas.

5. Integration of Clinical standards: CDSS can combine known clinical standards into their decision-making process. Clinical

standards are evidence-based suggestions created by expert groups to help healthcare workers in the treatment of particular diseases. By matching AI algorithms with clinical standards, CDSS can guarantee that the suggestions given are in line with the latest medical information and best practices. This combination improves diagnostic accuracy and enables obedience to regular care practices.

6. Personalization and Contextualization: AI systems in CDSS applying case-based reasoning and clinical standards can adapt suggestions based on individual patient traits. By considering patient-specific factors such as illnesses, drug history, and personal information, CDSS can tailor its ideas to the unique needs of each patient. This personalized method improves diagnosis accuracy and treatment planning, accounting for individual differences and improving patient care.

The merging of case-based thinking and clinical standards in CDSS uses the power of AI systems to help healthcare workers in making accurate and informed decisions. By tapping the vast knowledge saved in the case database and matching with evidence-based clinical standards, CDSS provides useful insights and suggestions, improving diagnostic accuracy and promoting optimal patient results. Continued breakthroughs in AI research, knowledge modeling, and case-based reasoning methods will further enhance the powers of CDSS, changing clinical decision-making in healthcare.

2.4.2.3 Real-Time Decision Support

Real-time decision support is a key component of clinical decision support systems (CDSS) that combine artificial intelligence (AI) tools. By deploying AI's powers, CDSS can provide fast and context-specific suggestions to healthcare workers, boosting diagnosis accuracy and improving patient results. Let's look into the merging of real-time decision support in CDSS and its effect on clinical decision-making.

1. Real-Time Data study: Real-time decision support in CDSS involves the study of patient data as it becomes available. AI systems can handle and analyze live data from various sources, such as electronic health records, vital signs monitors, smart devices, and test results. By constantly watching and studying this real-time data, CDSS can provide up-to-date insights and suggestions to healthcare workers, allowing quick actions and improving diagnostic accuracy.

2. Immediate Alerts and messages: CDSS with real-time decision support can generate immediate alerts and messages for important events or odd results. AI systems can spot trends or differences in the live data and cause alerts to warn healthcare workers of possible problems. These alerts can range from drug combinations and bad events to abnormal test results, spurring quick action and lowering the risk of diagnostic mistakes.

3. Clinical Decision Rules: Real-time decision support in CDSS includes set clinical decision rules, which are programs intended to guide specific clinical acts based on predefined criteria. These rules can be based on healthcare standards, best practices, or institutional protocols. By constantly tracking patient data, CDSS can apply these decision rules in real-time and provide context-specific suggestions to healthcare workers, ensuring obedience to standardized care routines.

4. Risk Assessment and Early action: Real-time decision help in CDSS allows early identification and action for high-risk patients. AI programs can study real-time data and patient trends to measure the chance of bad events or decline in health. By effectively warning healthcare workers to these risks, CDSS enables early intervention, quick treatment changes, and the prevention of avoidable problems.

5. Clinical process Integration: Real-time decision support in CDSS easily fits into the clinical process, giving suggestions at the point of care. CDSS can connect with electronic health record systems, clinical recording tools, and other healthcare information systems, allowing healthcare workers to receive decision support within their regular processes. This connection ensures that real-time suggestions are easily available and can be successfully integrated into professional decision-making processes.

6. Context-Specific Recommendations: Real-time decision help in CDSS considers the context of the patient's state and the clinical setting. By studying patient data in real-time, CDSS can provide context-specific suggestions suited to the individual patient's needs, clinical circumstances, and available resources. This personalized method improves diagnosis accuracy and treatment planning, helping healthcare workers in making well-informed choices matched with the real-time patient data.

7. Continuous Learning and Improvement: AI algorithms built into CDSS with real-time decision support can continuously learn and improve their suggestions based on feedback and results data. By studying the impact of decision support on patient results and adding comments from healthcare workers, CDSS can improve their models and enhance the accuracy and usefulness of real-time suggestions over time.

The inclusion of real-time decision support in CDSS provides healthcare workers with fast, context-specific suggestions, thereby improving diagnosis accuracy and patient results. By leveraging AI systems to examine real-time data, create alerts, apply clinical decision rules, and enable early action, CDSS improves clinical decision-making at the point of care. Continued study, evaluation, and teamwork between AI experts and healthcare workers will drive further improvements in real-time decision support, changing healthcare service and patient results.

AI Revolution in Medicine, The Future of Healthcare

2.5 Challenges and Ethical Considerations in AI Diagnostics

2.5.1 Data Quality and Bias

While the integration of artificial intelligence (AI) in diagnosis offers great promise, it also presents several obstacles and social concerns. One of the main challenges is ensuring the quality of data used to teach AI systems and the possible bias that may come from this data. Let's look into the importance of data quality and bias in AI diagnoses.

1. Data Quality: The accuracy and reliability of AI tests greatly depend on the quality of the data used to teach the algorithms. Issues such as missing or unclear data, mistakes in data entry, and data collection biases can affect the performance and dependability of AI systems. It is important to ensure that the training data is thorough, representative of various groups, and free from mistakes or biases to achieve maximum diagnostic accuracy.

2. Bias in Data: AI systems learn trends and make predictions based on the data they are taught on. If the training data includes biases, such as underrepresentation of certain categories or overrepresentation of specific population groups, the AI system may acquire those biases. This can lead to differences in diagnostic accuracy, as the algorithms may work differently for different patient groups. Addressing and reducing flaws in training data is important to ensure fairness and justice in AI diagnoses.

3. Generalization and External Validation: AI systems trained on specific datasets may not transfer well to new or different groups. The success of AI tests should be verified on external datasets that represent different patient groups and hospital settings. Robust evaluation ensures that the algorithms work correctly across various settings and lower the risk of biased or incorrect findings.

4. Transparency and Explainability: AI tests often involve complicated algorithms that can be challenging to understand and describe. Healthcare workers and patients need to understand how AI systems arrive at their medical choices. The lack of openness and explainability can hinder trust in AI diagnoses. Efforts should be made to build interpretable AI models and provide clear descriptions of the decision-making process to improve understanding and acceptance.

5. Accountability and Liability: The arrival of AI in testing brings questions of accountability and liability. In the case of medical mistakes or bad results, who should be held responsible—the AI system, the healthcare worker, or both? Clear rules and legal frameworks need to be created to describe the roles, responsibilities, and responsibility of all parties involved in AI tests to ensure patient safety and proper sharing of liability.

6. Patient Privacy and Consent: AI tests rely on access to patient data, which raises issues regarding patient privacy and data security. Strict processes must be in place to ensure the safe keeping, transfer, and anonymization of patient data. Informed consent should be gained from patients, clearly describing the goal and possible risks linked with the use of their data for AI tests. Protecting patient privacy and keeping secrecy are crucial ethical concerns.

7. Continuous Monitoring and Improvement: AI data should be continuously watched and improved to address new issues, biases, and mistakes. Regular review and confirmation of AI systems are important to ensure ongoing correctness, dependability, and fairness. Feedback methods should be created to collect data on medical results, spot areas of change, and update the models accordingly.

Addressing the hurdles and ethical factors in AI diagnoses, such as data quality and bias, is important to harness the full potential of

AI while ensuring fairness, accuracy, and patient trust. Striving for diverse and representative datasets, promoting transparency and explainability, defining accountability and liability frameworks, safeguarding patient privacy, and maintaining continuous monitoring and improvement will contribute to the responsible and ethical deployment of AI in diagnostics.

2.5.2 Transparency and Explainability

Transparency and explainability are crucial factors in the integration of artificial intelligence (AI) in diagnosis. As AI algorithms become increasingly complicated, it becomes important to understand and explain how these algorithms arrive at their medical choices. Let's study the importance of openness and explainability in AI diagnosis.

1. Building Trust and Acceptance: Transparency and explainability are important for building trust between healthcare workers, patients, and AI systems. When medical choices are made by AI systems, it is important for healthcare workers to understand the basic thinking and data behind those decisions. Similarly, patients should have access to details regarding how AI is utilized in their evaluation. Transparent and explainable AI tests can promote trust, improve acceptance, and empower healthcare workers and patients to make well-informed choices.

2. Interpretable AI Models: Interpretable AI models refer to programs that give results that are clear and explainable to people. In contrast to black-box models, which provide little input into the decision-making process, interpretable AI models provide reasons that can be understood and proven by healthcare experts. Developing and applying interpretable AI models in diagnostics allows healthcare workers to measure the dependability of the diagnostic results and spot possible biases or mistakes.

3. Providing reasons: AI systems should be capable of providing reasons for their medical choices. These explanations can take various forms, such as showing important features, presenting supporting evidence, or demonstrating the logical steps taken by the program. By giving clear and understandable answers, healthcare professionals can better understand the logic behind AI-generated findings, allowing effective teamwork between AI systems and healthcare workers.

4. Algorithmic Transparency: Transparency in AI research includes showing the inner workings of the algorithms and the data they are taught on. This openness helps healthcare workers to understand how the algorithms are created, the features they highlight, and any possible flaws that may exist. Transparent AI tests ensure that healthcare workers can assess the dependability and fairness of the diagnostic results, allowing them to make informed choices and act if necessary.

5. Ethical and Legal factors: Transparency and explainability in AI evaluations also address ethical and legal factors. Patients have the right to know and understand the steps involved in their evaluation. Moreover, healthcare workers must be able to explain and defend the medical choices made by AI systems to ensure responsibility and to meet legal and governmental requirements. Transparent and explainable AI tests add to ethical practices, governmental compliance, and the protection of patient rights.

6. ongoing Improvement and Learning: Transparency and explainability help the ongoing improvement and learning of AI systems. By offering insights into the decision-making process, healthcare workers can spot areas for growth, possible biases, and sources of mistakes. Transparent AI diagnoses allow for continuing improvement of algorithms, training data, and decision-making rules, ensuring that the systems grow and improve over time.

Promoting openness and explainability in AI investigations is important for promoting trust, ensuring responsibility, and improving the ethical use of AI in healthcare. By developing interpretable AI models, providing explanations for diagnostic decisions, embracing algorithmic transparency, and addressing ethical and legal considerations, healthcare professionals can effectively utilize AI systems as valuable tools in the diagnostic process, ultimately improving patient outcomes.

2.5.3 Legal and Regulatory Challenges

The inclusion of artificial intelligence (AI) in diagnosis brings about different legal and regulatory obstacles. As AI technology improves and becomes more common in healthcare, it is important to handle these issues to ensure patient safety, protect privacy, and keep ethical standards. Let's explore some of the key legal and regulatory factors in AI diagnosis.

1. Data Privacy and Security: AI treatments rely on access to patient data, including sensitive health information. Ensuring compliance with data protection laws, such as the General Data Protection Regulation (GDPR) or the Health Insurance Portability and Accountability Act (HIPAA), is important. Healthcare groups and AI makers must adopt strong security steps to protect patient data from breaches or illegal access.

2. Informed Consent: AI diagnostics may involve the use of patient data for teaching AI algorithms or proving their performance. Obtaining educated permission from people regarding the use of their data is important. Patients should be told about how their data will be used, the possible risks and rewards, and their rights regarding data privacy and security.

3. Liability and Accountability: Determining liability and accountability in AI research can be complicated. In situations where medical mistakes or bad results occur, it can be difficult to

attribute blame simply to the AI system or the healthcare worker. Establishing clear rules and legal models that outline the roles, responsibilities, and liabilities of all parties involved is important to ensure responsibility and patient safety.

4. Regulatory Approval and Oversight: AI systems used in tests may be called medical equipment and, therefore, subject to regulatory approval and oversight. Regulatory bodies, such as the Food and Drug Administration (FDA) in the United States, play a vital part in analyzing the safety, efficiency, and performance of AI-based detection systems. Ensuring that AI tests meet legal standards and undergo suitable testing and evaluation processes is important to protect patient safety.

5. Bias and Fairness: Addressing bias in AI diagnostics is a major law and regulatory issue. Biases can come from the data used to teach the algorithms or from the algorithmic design itself. Regulatory authorities are increasingly focused on ensuring fairness, openness, and responsibility in AI systems. Efforts are being made to create rules and standards that minimize bias and promote fair healthcare results.

6. Intellectual Property and Ownership: AI tests involve the building of algorithms and models that may be subject to intellectual property rights. Clarifying legal rights and licensing deals linked to AI diagnostics is crucial to promote creativity, incentivize research and development, and ensure proper use and spread of AI technologies.

7. Clinical Validation and Evidentiary Standards: Demonstrating the clinical validity and effectiveness of AI tests is important for legal permission and adoption in healthcare situations. Establishing suitable evidentiary standards and evaluation methods to evaluate the accuracy, dependability, and clinical effect of AI-based

diagnostic systems is important to ensure patient safety and drive evidence-based decision-making.

Addressing the legal and governmental issues associated with AI tests is important for their responsible and ethical application. Collaboration among lawmakers, regulatory authorities, healthcare workers, AI developers, and legal experts is important to create frameworks that protect patient rights, support fair and inclusive healthcare, and encourage innovation in AI investigations. Continuous tracking, adaption, and refining of laws and standards will be key to ensuring the safe and efficient use of AI in diagnoses.

2.5.4 Ethical Considerations in AI Diagnostics

The inclusion of artificial intelligence (AI) in diagnostics brings important social concerns that must be carefully handled. While AI has the potential to improve diagnosis accuracy and patient results, it is important to ensure that its use fits with social standards and protects the well-being of people. Let's study some key ethical concerns in AI diagnosis.

1. Beneficence and Non-Maleficence: The ethical concept of beneficence emphasizes the duty to act in the best interests of patients, while non-maleficence shows the importance of avoiding harm. AI diagnosis should value patient well-being by giving accurate and reliable findings. Measures must be in place to reduce the risk of medical mistakes and bad results, and to continuously watch and improve the performance of AI systems to ensure patient safety.

2. justice and equality: AI tests should be developed and applied in a way that ensures justice and equality for all people. It is important to handle flaws that may be present in the data used to teach AI systems. Efforts should be made to create varied and relevant records that cover different groups and communities.

AI Revolution in Medicine, The Future of Healthcare

Additionally, ongoing tracking and reporting of AI systems can help spot and correct any flaws that may appear in the diagnostic process.

3. liberty and Informed Consent: Respecting patient liberty and ensuring informed consent are important ethical factors in AI diagnosis. Patients should be told about the use of AI in their diagnosis, including its limits, possible benefits, and risks. Healthcare workers should engage in open and understandable dialogue to enable people to make informed choices regarding their healthcare. Patients should also have the choice to opt out of AI-driven testing methods if they so choose.

4. Privacy and secrecy: The use of AI in testing needs access to patient data, which raises issues regarding privacy and secrecy. Strict processes must be in place to protect patient information and ensure compliance with relevant privacy regulations. AI systems should stick to best practices in data anonymization, encryption, and safe keeping to protect patient privacy and security.

5. openness and Explainability: AI diagnosis should aim for openness and explainability. Healthcare workers and patients should be able to understand the decision-making process of AI systems and the reasoning behind the medical results. Transparent and explainable AI models can improve trust, allow teamwork between AI systems and healthcare experts, and support shared decision-making.

6. Professional Responsibility and Oversight: Healthcare workers have a responsibility to use AI diagnoses properly, legally, and within their area of practice. They should keep control and critical review of AI-driven medical choices, ensuring that these decisions match with professional opinion and the specific needs of individual patients. Professional groups and governing bodies play a vital part in giving rules and standards for the proper use of AI in testing.

7. Continuous tracking and Evaluation: Ethical factors in AI treatments require ongoing tracking and evaluation of AI systems. Regular review of the correctness, fairness, and performance of AI systems is important. Feedback systems should be created to collect data on diagnostic results, find areas of improvement, and address any ethics issues that may arise during the use of AI diagnostics.

Addressing ethical considerations in AI investigations requires a joint effort involving healthcare workers, AI developers, lawmakers, and patients. By supporting ethical principles such as beneficence, fairness, liberty, and privacy, and by promoting openness, informed consent, and constant tracking, the inclusion of AI in diagnosis can match with ethical standards and add to better patient care.

2.6 Future Perspectives and Impact of AI on Diagnostics

2.6.1 The Role of AI in Precision Medicine

AI is set to play a transformative role in the field of testing, especially in the context of precise medicine. Precision medicine aims to adapt medical care and solutions to individual patients based on their unique genetic, environmental, and lifestyle factors. Let's explore the role of AI in improving precise medicine and its possible effect on tests.

1. Personalized Risk Assessment: AI systems can examine vast amounts of data, including genetic information, health records, and lifestyle data, to find trends and signs linked with disease risk. By deploying AI methods, healthcare workers can perform personalized risk assessments, allowing early diagnosis and assistance for people who are at a higher risk of getting certain illnesses. AI-driven risk assessment models can provide more accurate and focused prevention tactics, eventually better patient results.

2. Predictive Modeling: AI has the ability to change the area of predictive modeling in testing. By studying patient data, including genetic profiles, medical records, and real-time tracking data, AI systems can spot minor trends and predict disease development with greater precision. Predictive models driven by AI can help healthcare workers expect disease results, improve treatment plans, and effectively engage to reduce risks.

3. Image-Based Phenotyping: AI systems, especially those applying deep learning methods, have shown great promise in image-based phenotyping. By studying medical pictures, such as imaging scans or pathological slides, AI can extract detailed details and spot minor features that may escape human viewers. This allows more accurate and objective measurement of disease traits, helping in early discovery, grading, and tracking of illnesses such as cancer or neurological disorders.

4. Treatment Response Prediction: AI systems can study complicated data, including patient traits, genetic profiles, and treatment records, to predict individual response to specific medicines. By considering various factors and utilizing machine learning techniques, AI can help healthcare workers in finding the most effective treatment choices for individual patients. This can lead to more personalized and focused actions, lowering the chance of harmful effects and improving treatment results.

5. Clinical Decision Support: AI-based clinical decision support systems (CDSS) can help healthcare workers in making more knowledgeable and evidence-based diagnosis choices. By studying patient data, symptoms, medical records, and relevant research papers, AI can provide customized tips and ideas to help healthcare workers in the diagnosis process. CDSS driven by AI can improve diagnostic accuracy, reduce diagnostic mistakes, and allow prompt actions.

6. Real-Time Monitoring and Surveillance: AI technologies allow real-time monitoring and surveillance of patients, continuously analyzing data streams from smart devices, sensors, or electronic health records. This allows for early discovery of errors or changes in health factors, allowing quick actions. AI-powered tracking systems can enhance disease control, improve patient commitment to treatment routines, and simplify online patient care.

7. Data merging and Knowledge Discovery: AI can enable the merging of diverse data sources, including genetics, electronic health records, and scientific books. By deploying AI methods, healthcare workers can discover secret patterns, connections, and insights from vast amounts of data. This knowledge finding process can fuel medical study, improve understanding of disease causes, and guide diagnosis methods.

The merging of AI in precision medicine holds great potential for changing diagnosis. By enabling personalized risk assessment, predictive modeling, image-based phenotyping, treatment response prediction, clinical decision support, real-time monitoring, and data integration, AI has the potential to enhance diagnostic accuracy, enable targeted interventions, and improve patient outcomes. Embracing AI technologies in testing can pave the way for a more accurate, efficient, and patient-centered approach to healthcare.

2.6.2 Collaborative Approaches: Human-AI Integration

As AI continues to improve in diagnoses, a joint method that combines the powers of AI with the knowledge of healthcare workers becomes increasingly important. Human-AI collaboration aims to leverage the skills of both people and AI systems to enhance diagnosis processes and improve patient care. Let's study the joint methods and perks of human-AI merging in testing.

1. enhanced Intelligence: Human-AI interaction promotes the idea of enhanced intelligence, where AI systems improve the skills

of healthcare workers rather than removing them. AI systems can handle vast amounts of data, spot trends, and provide insights to help medical decision-making. By working with AI, healthcare workers can leverage this wealth of knowledge to make more accurate and informed assessments.

2. Data Analysis and Interpretation: AI excels at analyzing and understanding large amounts of complex data, such as genetic data or medical imaging scans. By harnessing AI tools, healthcare workers can obtain faster and more thorough studies, allowing them to focus on understanding and applying the results to patient care. AI can help in finding critical features, trends, or errors in data, allowing more precise evaluations and treatment suggestions.

3. Clinical Decision Support: AI can serve as a useful clinical decision support tool, giving healthcare workers with evidence-based suggestions and reports based on patient data and medical standards. By adding AI-driven decision support systems into diagnostic processes, healthcare workers can benefit from fast and relevant information, lowering the risk of diagnosis mistakes and improving patient safety.

4. Enhanced Efficiency and Productivity: Human-AI merging in testing can greatly improve efficiency and productivity. AI systems can process and study data at a much faster rate than people, allowing for shorter response times in medical processes. This efficiency gain helps healthcare workers to focus more on patient contact, care planning, and other complicated tasks that require their unique skills and knowledge.

5. Quality Assurance and Validation: Human monitoring and validation are crucial in ensuring the accuracy and trustworthiness of AI-driven diagnosis systems. Healthcare workers play a key role in confirming AI outputs, checking diagnoses, and closely analyzing the practical usefulness of AI-generated information. Human

knowledge is important for ensuring that AI systems match with the unique needs and traits of individual patients.

6. Ethical and Social factors: Human-AI interaction offers a chance to solve ethical and social factors in diagnosis. Healthcare workers can add their ethical reasoning, understanding, and compassion to ensure that patient values, tastes, and varied ethnic backgrounds are considered in the diagnosis process. By mixing AI's logical powers with human values and sensitivity, diagnoses can be more patient-centered and socially sound.

7. ongoing Learning and Adaptation: Human-AI teamwork allows for ongoing learning and improvement in diagnosis. As healthcare workers work with AI systems, they can provide input, confirm AI results, and add to improving and updating algorithms. This iterative process allows AI systems to learn from real-world clinical experiences, adapt to changing healthcare dynamics, and continually improve their diagnostic capabilities.

The joint merging of AI and healthcare experts in diagnosis has the potential to change patient care. By combining the strengths of both people and AI systems, medical accuracy can be improved, speed can be enhanced, and patient results can be maximized. Embracing a joint approach to human-AI integration in testing holds great potential for improving the field and offering high-quality, patient-centered healthcare.

2.6.3 Enhancing Accessibility and Affordability of Diagnostics

The integration of AI in diagnostics has the potential to handle the issues of accessibility and price, making diagnostic services more widely available and cost-effective. Let's study how AI can improve usability and cost in testing.

1. rural and Point-of-Care Diagnostics: AI-powered diagnostic tools can be applied in rural and underdeveloped areas, bringing

diagnostic skills to regions with limited access to healthcare centers. With the help of movable devices and AI algorithms, healthcare workers can perform tests at the point of care, lowering the need for patients to drive long distances or wait for extended times to receive diagnostic services. This allows early discovery, prompt action, and better patient outcomes, especially in resource-constrained situations.

2. Telemedicine and Virtual appointments: AI technologies can support telemedicine efforts by enabling remote medical appointments. Through video chat and online access to patient data, healthcare workers can interact with AI systems to evaluate signs, medical records, and diagnosis results. AI programs can help in triaging patients, offering basic diagnoses, and leading treatment suggestions. Telemedicine driven by AI can expand access to medical knowledge, especially for individuals living in rural or impoverished areas.

3. Streamlined Workflows and Efficiency: AI technologies can improve diagnostic workflows, lowering response times and improving the efficiency of diagnostic processes. AI programs can handle regular jobs, such as data entry, picture analysis, and report creation, allowing healthcare workers to focus on critical aspects of evaluation and treatment planning. By simplifying processes, AI can help healthcare centers handle a larger amount of diagnostic cases, better access to diagnostic services for a greater number of patients.

4. Cost Reduction: AI-driven tests have the ability to lower costs connected with diagnostic processes. By handling chores and improving efficiency, AI can reduce the need for human work and resources. Furthermore, AI can help in finding the most appropriate medical tests, avoiding needless and costly treatments. The possible cost cuts made through AI diagnosis can make these services more cheap and available to a bigger audience.

AI Revolution in Medicine, The Future of Healthcare

5. Preventive and Early Detection Strategies: AI systems can play a major part in preventive and early detection strategies, leading to better health results and lower healthcare costs. By leveraging AI-driven risk assessment models, healthcare workers can spot people at higher risk of getting certain illnesses and apply focused protective measures. Early discovery of diseases through AI-powered screening programs can allow quick treatments, lowering the need for expensive and invasive testing methods at later times.

6. Scalability and Reproducibility: AI-powered diagnostic systems can be quickly scaled and copied across different healthcare situations, offering uniform and regular diagnostic services. This flexibility allows the spread of diagnostic knowledge and best practices, lowering regional gaps in access to quality diagnostics. Moreover, AI systems can continually learn and improve from a large pool of varied data, adding to the accuracy and generalizability of diagnosis results.

7. personalized and adaptable Diagnostics: AI technologies can allow personalized and adaptable diagnosis methods suited to individual patient needs. By combining patient-specific data, such as genetic profiles, living information, and medical background, AI systems can provide unique diagnostic insights. This customization improves diagnostic accuracy, lowers needless tests and processes, and ensures that patients receive focused and cost-effective diagnostic services.

The inclusion of AI in diagnostics holds vast potential to improve the usability and cost of diagnostic services. By enabling remote diagnostics, optimizing workflows, reducing costs, facilitating preventive strategies, and ensuring scalability and customization, AI-driven diagnostics can democratize access to high-quality diagnostic services, improve patient outcomes, and contribute to more equitable healthcare delivery.

AI Revolution in Medicine, The Future of Healthcare

AI Revolution in Medicine, The Future of Healthcare

CHAPTER 3: AI IN TREATMENT AND PERSONALIZED MEDICINE

3.1 The Paradigm Shift towards Personalized Medicine

3.1.1 Understanding Personalized Medicine

In recent years, there has been a major change in healthcare towards personalized medicine, also known as precision medicine. This method understands that each patient is unique, and their healthcare needs can be better handled by fitting treatments to their individual traits, genetic makeup, lifestyle, and external factors. Let's dig into learning personalized medicine in more detail.

Personalized medicine aims to move away from the standard one-size-fits-all method to healthcare. It recognizes that people with the same disease may respond differently to treatments based on their individual differences. By considering individual factors, tailored medicine tries to optimize treatment results, reduce harmful effects, and improve total patient well-being.

Key elements of specialized medicine include:

1. genomes and Molecular Profiling: Personalized medicine puts a strong focus on genomes and molecular profiling. Advances in genome sequencing technologies have made it possible to study an individual's genetic makeup and find specific genetic differences that may affect their disease risk, drug metabolism, and treatment response. Molecular profiling methods also measure other biomarkers, such as proteins or chemicals, to gain a complete understanding of a patient's unique molecular features.

2. Tailored Treatment Strategies: Based on the patient's genetic and biological makeup, personalized medicine allows the

development of tailored treatment strategies. Treatment choices can be led by genetic markers that suggest a patient's chance of reacting to a particular treatment, possible drug combinations, or higher risk of bad effects. This method allows healthcare workers to pick the most effective and best treatment choices for each patient.

3. Predictive and preventative Medicine: Personalized medicine combines predictive and preventative medicine methods. By studying DNA and cellular data, healthcare workers can spot people who are at a higher risk of getting certain illnesses. Early treatments, such as lifestyle changes or focused tests, can be adopted to avoid or identify illnesses at an early stage when they are more manageable. This proactive method helps to improve patient results and lower healthcare costs involved with treating severe illnesses.

4. Integration of Patient Data: Personalized medicine combines different types of patient data, including clinical data, genetic information, lifestyle factors, environmental risks, and real-time tracking data. By joining these data sources, healthcare workers can gain a complete view of the patient's health and make more informed choices regarding treatment selection and disease control. Advanced data analytics and AI play a vital part in studying and understanding this vast amount of data.

5. Patient-Centric Care: Personalized medicine puts the patient at the center of care. It sees the importance of involving patients in treatment choices and fitting methods to their personal needs and desires. Informed permission, joint decision-making, and patient education are important components of tailored medicine, enabling patients to actively participate in their own healthcare path.

Personalized medicine has the ability to change healthcare by improving treatment results, lowering side effects, and increasing resource usage. The combination of AI and advanced technologies further improves the powers of personalized medicine by allowing the analysis of vast amounts of data and producing insights for personalized treatment strategies. As study and technology advances continue, personalized medicine is expected to play an increasingly significant role in providing focused and effective healthcare treatments to people.

3.1.2 Benefits and Challenges of Personalized Medicine

Personalized medicine brings a range of benefits and possibilities to healthcare, but it also offers unique challenges. Let's review the rewards and difficulties connected with personalized health.

Benefits of Personalized Medicine:

1. Improved Treatment Outcomes: By customizing treatments to individual patients based on their genetic, molecular, and clinical traits, tailored medicine aims to improve treatment success. This method can lead to improved patient results, including higher reaction rates, better disease control, and greater mortality rates.

2. Reduced Adverse Effects: Personalized medicine considers individual differences in drug processing and reaction, reducing the chance of adverse effects. By avoiding treatments that are unlikely to be successful or are linked with a higher risk of side effects, personalized medicine increases patient safety and improves tolerance of therapies.

3. Optimized Resource Utilization: Personalized medicine can help optimize the use of healthcare resources by targeting treatments to those who are most likely to benefit. By finding patients who are unlikely to react to certain medicines, needless treatments can be avoided, saving costs and resources for more suitable interventions.

4. Early Detection and avoidance: Personalized medicine stresses the early detection and avoidance of illnesses. Genetic and molecular testing can identify people at higher risk of getting specific diseases, allowing for proactive treatments, lifestyle

changes, and focused tests. Early diagnosis and protection measures can lead to better health results and lower healthcare costs.

5. Targeted treatments: Personalized medicine allows the development of targeted treatments that address the individual genetic flaws causing a patient's illness. By knowing the basic processes of a disease at an individual level, medicines can be developed to accurately target those abnormalities, improving treatment effectiveness and lowering off-target effects.

Challenges of Personalized Medicine:

1. Data Complexity and Integration: Personalized medicine depends on the integration and analysis of diverse data sources, including genetic data, health records, lifestyle information, and environmental factors. Managing and understanding this complicated and vast amount of data can be difficult, requiring strong infrastructure, advanced data analytics, and skill in data analysis.

2. Ethical and Privacy Considerations: Personalized medicine raises ethical issues linked to the use and keeping of private DNA and health information. Ensuring patient privacy, getting informed permission, and keeping data security are crucial aspects that need to be handled to build trust and protect patient rights in personalized medicine efforts.

3. Cost and Accessibility: The application of personalized medicine methods can be expensive, requiring genetic tests, molecular analysis, and focused treatments. Ensuring cost and usability of personalized medicine for all patients, regardless of financial position, continues a problem. Balancing the possible benefits with the costs involved with personalized medicine is important for fair healthcare delivery.

4. Limited proof and Validation: Some personalized medicine methods are relatively new and may have limited proof supporting their clinical usefulness. Robust confirmation studies and clinical trials are important to prove the efficiency, safety, and cost-effectiveness of personalized medicine treatments. Generating sufficient proof to back the adoption of personalized medicine methods can be time-consuming and difficult.

5. Integration into Clinical Practice: Integrating personalized medicine methods into regular clinical practice can be complicated. Healthcare workers need proper training, equipment, and decision support tools to successfully adopt personalized medicine tactics. Overcoming obstacles linked to process integration, payment models, and physician acceptance is important for successful integration into healthcare systems.

Despite the challenges, tailored medicine holds great potential to change healthcare by tailoring treatments to individual patients, better treatment results, and increasing resource usage. Continued study, technological breakthroughs, and joint efforts are important to beat the challenges and fully achieve the benefits of personalized medicine in better patient care.

3.2 AI and Drug Discovery

3.2.1 Accelerating Drug Discovery Process with AI

The process of finding and creating new drugs is complicated, time-consuming, and expensive. However, AI has appeared as a powerful tool in speeding the drug development process, changing the way new medicines are discovered and created. Let's explore how AI is changing drug development and speeding the search for new treatments.

1. Target Identification and Validation: AI systems can examine vast amounts of biological data, including genetic data, protein

interactions, and disease processes, to find possible drug targets. By integrating data from various sources and applying machine learning techniques, AI can predict the chance of a target being medically useful. This allows researchers to select targets with the best possibility for good drug creation, saving time and resources.

2. Virtual Screening and Drug Design: AI can enable virtual screening, which includes using computer models to screen big databases of chemicals and predict their potential for sticking to specific targets. Machine learning systems can learn from known drug-target interaction data to predict the binding and specificity of new molecules. This speeds the recognition of potential drug prospects, lowering the need for lengthy laboratory tests.

3. De Novo Drug creation: AI methods, such as generative models and deep learning, allow the de novo creation of new drug-like molecules. These models learn patterns from known chemicals and create new molecules with desired qualities, such as high efficiency and low toxicity. AI-driven de novo drug creation widens the chemistry space for study, providing researchers with a wider range of possible drug options to assess.

4. Predictive Analytics and Drug Repurposing: AI systems can examine large-scale data sets, including clinical data, genome data, and drug databases, to predict drug reactions and reuse known drugs for new purposes. By finding trends and connections, AI can discover drugs that may have effectiveness against different diseases or provide alternative treatment choices. This method greatly shortens the drug research timeframe and lowers costs.

5. Optimization of Clinical Trials: AI can optimize clinical trial design by predicting patient reactions, finding suitable patient groups, and improving dose methods. By combining various data sources, including patient traits, genetic information, and real-world

proof, AI programs can improve trial efficiency and increase the chance of success.

6. Safety Assessment and Toxicity Prediction: AI models can predict the safety and toxicity profiles of drug options, helping in the early identification of possible harmful effects. By studying chemical structures and known toxicity data, AI systems can flag compounds with a higher chance of safety problems, allowing researchers to select safer options for further development.

The use of AI in drug development offers several benefits, including faster identification of drug targets, more efficient screening of chemicals, and higher success rates in clinical studies. AI-driven methods can greatly reduce the time and cost involved in the drug development process, allowing researchers to study a wider range of possible treatments and speed the release of new medicines to patients in need.

However, hurdles continue, including the need for high-quality data, interpretability of AI models, and legal concerns. Collaborative efforts between researchers, AI experts, and regulatory bodies are important to handle these challenges and ensure the safe and effective merging of AI in drug development. As AI technologies continue to improve, they hold the promise of changing the drug finding environment, leading to more focused, effective, and personalized treatments for a wide range of illnesses.

3.2.2 AI in Target Identification and Validation

Target identification and validation are important steps in the drug development process, and AI has emerged as a useful tool in speeding and improving these stages. AI methods allow researchers to examine vast amounts of data, find possible drug targets, and confirm their therapeutic usefulness. Let's study how AI is changing target recognition and proof in drug development.

1. Data Integration and Analysis: AI systems can combine diverse data sources, including genetic data, protein-protein interactions, book databases, and clinical data, to find possible drug targets. By studying this wealth of information, AI can reveal secret patterns, connections, and possible drug-target interactions that may not be instantly clear to human researchers. This data-driven method allows the discovery of new targets and improves the knowledge of complicated illnesses.

2. Prediction of Therapeutic Relevance: AI models, such as machine learning and deep learning algorithms, can examine large-scale biology and clinical data sets to predict the therapeutic relevance of possible drug targets. By learning from known drug-target interactions and disease-related data, AI systems can make predictions on the chance of a target being involved in a specific disease process or having possible treatment value. This helps researchers to select targets with a higher chance of success, lowering the time and cost involved with trial confirmation.

3. Virtual Screening and Drug Repurposing: AI-powered virtual screening methods allow the fast screening of big chemical libraries against possible drug targets. Machine learning systems can learn from known drug-target interaction data and predict the binding affinity and specificity of new chemicals. This allows researchers to find good drug prospects and adapt known drugs for new purposes, possibly saving years of development time.

4. Network Analysis and Pathway Exploration: AI programs can study complicated biological networks, such as protein-protein interaction networks or signaling pathways, to find key hubs or dysregulated pathways that may serve as possible drug targets. By finding links and identifying key biological processes, AI helps researchers understand the basic mechanisms of illnesses and discover new paths for therapeutic intervention.

5. merging of Omics Data: AI methods enable the merging and study of multi-omics data, including genomes, transcriptomics, proteomics, and metabolomics. By joining these layers of knowledge, AI systems can find possible targets that are dysregulated at different levels and gain a more complete picture of illness processes. This collaborative method improves the accuracy and dependability of target recognition and evaluation.

6. High-Throughput Experimental Design: AI systems can guide the design of high-throughput tests to confirm possible drug targets. By improving testing settings and sample selection, AI helps researchers produce reliable and repeatable data, increasing the effectiveness of target validation studies. This data-driven approach reduces the reliance on traditional trial-and-error methods and speeds the discovery of potential targets.

The application of AI in target recognition and validation offers several benefits, including the ability to study large and complex data sets, reveal secret connections, and predict treatment usefulness. AI-driven methods improve the efficiency, accuracy, and success rates of target identification, eventually speeding up the drug development process and increasing the chance of finding successful treatments for different illnesses.

However, obstacles remain, such as the quality and quantity of data, interpretability of AI models, and the need for scientific confirmation. Collaborative efforts among academics, bioinformaticians, and AI experts are vital to solve these challenges and ensure the reliable and responsible use of AI in target selection and validation. Continued improvements in AI technologies, along with diverse cooperation, hold great promise for improving the finding of new drug targets and advancing the creation of innovative treatments.

3.2.3 AI in Drug Repurposing and Combination Therapy

AI plays a major role in speeding drug development not only through target recognition but also in the areas of drug reuse and combination treatment. Leveraging AI methods, researchers can study current drugs for new purposes and find synergistic mixtures of drugs to improve treatment effectiveness. Let's look into how AI is changing drug reuse and allowing the development of combo treatments.

1. Drug Repurposing: a. Data Mining and Integration: AI programs can examine vast amounts of data, including electronic health records, genetic data, drug databases, and scientific books, to find possible possibilities for drug reuse. By finding patterns and connections, AI can suggest existing drugs that may have healing benefits in different diseases or conditions beyond their original indications. b. Computational Predictions: AI models can predict the efficacy of reused drugs by leveraging machine learning techniques and integrating diverse data sources. These models learn from known drug-target interactions and biological processes to measure the chance of a drug being useful against a particular target or disease. This computer method expedites the discovery of possible reuse candidates and lowers the need for thorough field testing. c. Safety and Side Effect Prediction: AI systems can analyze the safety and side effect patterns of reused drugs by studying chemical structures, known harmful effects, and patient data. By predicting possible safety issues, AI helps researchers find drugs with a positive benefit-risk ratio for recycling, saving time and resources in the drug creation process.

2. Combination Therapy: a. Synergy Prediction: AI models can examine large-scale data sets, including genetic data, drug-target interactions, and clinical data, to predict beneficial interactions between different drugs. By finding similar modes of action and possible drug-drug interactions, AI helps researchers discover new mixtures that can improve treatment effectiveness and beat drug

resistance. b. specialized Combination Therapy: AI systems can examine patient-specific data, such as genetic profiles, disease traits, and therapy experiences, to suggest specialized combination medicines. By considering individual differences, AI can improve treatment methods tailored to each patient's unique needs, increasing the chances of therapy success. c. Drug Interaction and Side Effect Assessment: AI can help in reviewing possible drug interactions and predicting harmful effects in combination treatments. By studying known drug-drug interactions, chemical structures, and patient data, AI systems can help researchers spot possible risks and create better combination treatment methods.

The use of AI in drug reuse and combination therapy offers several benefits, including the possibility for faster discovery of new applications, reduced costs compared to de novo drug research, and the ability to improve treatment results by mixing beneficial therapies. AI-driven methods open up new options for studying current drugs and tailoring treatments to individual patients.

However, obstacles remain, such as the need for high-quality data, understanding of AI models, and the requirement for thorough clinical confirmation. Collaborative efforts among researchers, doctors, and AI experts are important to handle these challenges and ensure the safe and effective application of AI in drug reuse and combination treatment.

By tapping the power of AI, researchers can find new therapeutic possibilities, adapt existing drugs, and improve treatment routines, eventually leading to more efficient and effective therapeutic approaches for a wide range of illnesses.

3.3 Precision Treatment Selection with AI
3.3.1 Precision Oncology: Tailoring Cancer Treatment

Precision oncology, also known as tailored or focused medicine, tries to adjust cancer treatment based on an individual's unique genetic, molecular, and clinical traits. AI plays a crucial part in precision cancer by studying large-scale data, predicting treatment outcomes, and leading therapeutic decision-making. Let's explore how AI is changing precision oncology and revolutionizing cancer treatment selection.

1. Genomic Data Analysis: a. Tumor Genomic Profiling: AI systems can study genomic data from tumor samples to find specific genetic changes, such as mutations, gene amplifications, and rearrangements. By matching the genetic makeup of the tumor with known cancer-associated genes and pathways, AI can determine possible treatment targets and guide therapy choices. b. Variant Classification: AI models can predict the functional importance of genetic variants identified in tumors, helping doctors distinguish between driver mutations (mutations that cause cancer growth) and passenger mutations (neutral mutations). This knowledge is important for choosing focused treatments that directly address the driver genes causing tumor growth. c. Biomarker Identification: AI methods, such as machine learning and deep learning, can study genetic and molecular data to find possible biomarkers associated with drug reaction or tolerance. By finding these signals, AI helps doctors identify patients who are likely to benefit from specific medicines and avoid needless treatments for those who are unlikely to react.

2. Treatment Response Prediction: a. Predictive Modeling: AI models can learn from big datasets that include patient traits, genetic profiles, treatment results, and clinical data to predict individual patient reactions to different treatment choices. By considering multiple factors and learning from past treatment response data, AI systems can create individual predictions of treatment success, helping doctors in choosing the most appropriate medicines for their patients. b. Clinical Decision Support: AI-driven clinical decision support tools can provide evidence-based treatment suggestions by

studying patient-specific data in real-time. These systems consider a patient's clinical features, cancer stage, biomarker state, and treatment recommendations, helping doctors in making informed treatment choices that fit with the principles of precision oncology.

3. Treatment Optimization and Drug Combinations: a. Combinatorial Treatment Optimization: AI systems can examine big datasets on drug interactions, biological pathways, and treatment reactions to find best combos of tailored therapies, immunotherapies, or chemotherapy agents. By considering the connection between different treatments and the unique traits of each patient, AI helps create custom combination medicines that improve treatment effectiveness and beat drug resistance. b. Adaptive Treatment Strategies: AI-driven models can flexibly change treatment strategies based on real-time patient data and treatment reactions. By constantly tracking and studying patient data, AI programs can suggest treatment tweaks, dose adjustments, or changes in therapy to improve treatment results and adapt to changing disease conditions.

4. Clinical Trial Matching: a. Patient Eligibility Assessment: AI systems can assess patient traits, disease factors, DNA data, and treatment history to match qualified patients with relevant clinical studies. This improves access to new treatments and ensures that patients have the chance to join in cutting-edge research. b. Trial Design Optimization: AI methods can help improve clinical trial design by studying past trial data, finding possible errors, and offering insights to enhance trial efficiency and patient recruitment.

The merger of AI in precision cancer offers numerous benefits, including better treatment results, fewer side effects, and enhanced patient care. However, difficulties continue, such as the need for high-quality data, interpretability of AI models, and legal concerns. Collaboration between researchers, doctors, data scientists, and regulatory bodies is vital to handle these issues and ensure the responsible and effective application of AI in precision cancer.

By tapping the power of AI, precision oncology holds the potential to change cancer treatment by offering individualized medicines that target the unique traits of each patient's tumor. This method can improve treatment response rates, increase survival, and boost the general quality of life for cancer patients.

3.3.2 AI in Predicting Treatment Response and Resistance

AI has emerged as a strong tool in identifying drug reaction and resistance in different diseases, including cancer. By studying complicated records and finding trends, AI systems can provide useful insights into patient-specific treatment results. Let's study how AI is changing the prediction of drug reaction and resistance.

1. Integration of Multimodal Data: a. genetic Data: AI systems can evaluate genetic data, including somatic mutations, gene expression profiles, and copy number changes, to predict treatment reaction. By linking specific genetic changes with drug sensitivity or resistance, AI can find genome signs that suggest the chance of a good reaction to a particular treatment. b. Clinical Data: AI models can combine clinical data such as patient traits, disease features, treatment history, and test reports to predict treatment reaction. By considering factors such as age, illnesses, and initial health state, AI systems can provide specific estimates of treatment results. c. Imaging Data: AI methods, including deep learning, can examine medical imaging data such as radiographic pictures, disease slides, or molecular imaging data. By pulling features and trends from these pictures, AI can predict treatment reaction and find areas of interest that may affect therapy results.

2. Machine Learning Models for Prediction: a. Predictive Modeling: AI models can learn from big files holding patient data, treatment plans, and treatment results to predict individual treatment

reaction. These models can find predictive factors and create individual predictions of treatment effectiveness, helping doctors in choosing the most appropriate medicines for their patients. b. Biomarker Identification: AI systems can examine high-dimensional omics data, such as genetics, transcriptomics, proteomics, and metabolomics, to find possible biomarkers associated with treatment reaction or tolerance. By finding these signals, AI helps doctors stratify patients into different subgroups and tailor treatment methods appropriately. c. Time Series Analysis: AI methods can study continuous patient data to predict treatment reaction over time. By tracking changes in biomarkers, clinical parameters, or imaging features, AI systems can provide real-time predictions of treatment effectiveness and allow quick adjustments in therapeutic interventions.

3. Drug Resistance Prediction: a. Resistance Mechanism Analysis: AI systems can examine molecular paths, drug-target relationships, and gene expression data to find possible causes of treatment resistance. By knowing the underlying biological processes, AI can predict the probability of building tolerance and guide the selection of alternative treatment methods. b. Dynamic Treatment Response Monitoring: AI-driven models can constantly watch treatment responses and spot early signs of resistance. By studying continuous patient data, AI systems can identify changes in biomarker expression or disease development patterns, allowing doctors to modify treatment plans before resistance becomes clinically apparent.

4. Clinical Decision Support Systems: a. Treatment suggestions: AI-driven clinical decision support tools can provide evidence-based treatment suggestions based on patient-specific data. By combining various data sources, including clinical data, genetic profiles, and treatment standards, AI systems can help doctors in making informed choices about treatment selection and improve

treatment results. b. Treatment Optimization: AI models can offer treatment improvements, dose tweaks, or changes in therapy based on real-time patient data. By considering individual patient traits and treatment reaction patterns, AI systems can change treatment plans to maximize efficiency and reduce resistance.

The application of AI in predicting treatment reaction and resistance holds great promise in leading personalized treatment choices. However, obstacles such as data quality, interpretability of AI models, and clinical proof need to be handled for broad usage. Collaborative efforts among researchers, doctors, and AI experts are crucial to building solid and dependable AI models that can successfully predict treatment results and improve patient care.

By tapping the power of AI, healthcare workers can make more informed treatment choices, pick the most suited methods for individual patients, and eventually improve treatment results in personalized medicine

3.3.3 Pharmacogenomics: Optimizing Drug Selection with AI

Pharmacogenomics is the study of how an individual's genetic makeup affects their reaction to medicines. By studying genetic differences, AI systems can provide useful insights into drug processing, effectiveness, and possible bad responses. Let's study how AI is changing drug selection and improvement through pharmacogenomics.

1. Genomic Data Analysis: a. Genetic Variations: AI systems can examine genome data, including single nucleotide variants (SNPs) and copy number variations (CNVs), to find genetic variations that affect drug reaction. By matching an individual's genetic makeup with known drug-gene interactions, AI can guess how a person may process or react to specific medicines. b. Pharmacokinetics and Pharmacodynamics: AI models can combine genetic data with information on drug pharmacokinetics (how the body processes the

drug) and pharmacodynamics (how the drug affects the body) to provide personalized estimates of drug reaction. By considering factors such as enzyme activity, drug transporter function, and receptor interactions, AI systems can improve drug selection based on an individual's genetic background.

2. AI-Driven Drug Selection: a. Treatment Response Prediction: AI systems can examine genetic data, clinical factors, and treatment results to predict the chance of response to specific medicines. By considering genetic differences linked with drug digestion and effectiveness, AI can help doctors in choosing the most suitable medicine for an individual patient. b. bad Drug Reaction Prediction: AI models can find genetic markers associated with a higher chance of bad drug responses. By studying a patient's genetic background, AI programs can help spot possible safety issues and guide the selection of alternative medicines or dose changes to reduce the risk of bad events.

3. Clinical Decision Support Systems: a. Integration of genetic Data: AI-driven clinical decision support systems can combine genetic data with patient-specific information, such as medical background, co-morbidities, and associated medicines. By considering DNA differences and known drug-gene interactions, these systems can provide specific suggestions for drug selection and doses. b. Treatment Guidelines and Evidence-Based suggestions: AI systems can examine large-scale clinical data, treatment guidelines, and drug databases to provide evidence-based suggestions for drug selection. By considering the latest study and real-world data, AI-driven systems can help doctors in making informed decisions and better patient results.

4. Real-Time Genomic Data Analysis: a. Point-of-Care Genomic Analysis: AI models can study genomic data in real-time, allowing fast identification of genetic differences that affect drug reaction. This helps doctors to make rapid treatment changes based on a

patient's genetic makeup. b. Integration with Electronic Health Records (EHRs): AI systems can combine genomic data with EHRs, giving doctors with complete patient profiles that include genetic information. This allows more informed decision-making and enables specific drug selection based on an individual's genetic traits.

The merging of AI in pharmacogenomics holds great potential in improving drug selection and avoiding harmful drug responses. However, obstacles such as data protection, legal factors, and clinical proof need to be handled for broad application. Collaboration among researchers, doctors, geneticists, and AI experts is important to develop solid AI models and set standards for bringing pharmacogenomics into clinical practice.

By tapping the power of AI in pharmacogenomics, healthcare workers can individualize drug therapy, increase treatment effectiveness, and reduce the risk of bad responses, eventually leading to better patient results and individualized medicine.

3.4 AI-Assisted Treatment Planning and Decision-Making

3.4.1 AI in Treatment Planning and Dosage Optimization

AI-assisted treatment planning and dose optimization have the potential to change healthcare by offering personalized and optimized treatment plans for patients. Let's study how AI is being used in treatment planning and dose improvement.

1. Treatment Planning: a. specialized Treatment Selection: AI systems can examine patient-specific data, including medical history, genetic information, biomarker readings, and treatment results, to suggest specialized treatment choices. By considering individual patient traits and the latest evidence-based standards, AI can help healthcare workers in choosing the most effective

treatments tailored to each patient's needs. b. Clinical Decision Support Systems: AI-driven clinical decision support systems can combine patient data, treatment standards, and medical books to provide evidence-based suggestions for treatment plans. By tapping AI's ability to study vast amounts of data and spot trends, these systems can help doctors in making informed choices and building complete treatment plans.

2. Dosage Optimization: a. unique Dosage Calculation: AI programs can consider various factors, such as patient data, genetic differences, disease features, and treatment response, to calculate unique drug amounts. By taking into account factors that affect drug digestion and clearance, AI can adjust amounts to achieve desired treatment results while reducing the risk of bad responses. b. Real-Time Monitoring and Adjustment: AI-assisted systems can watch patients' vital data, treatment reaction, and bad events in real time. By studying this data, AI programs can provide suggestions for dose changes or treatment modifications to ensure optimal effectiveness and safety.

3. Predictive Analytics: a. Treatment Response Prediction: AI models can examine patient data, including clinical factors, genetic information, biomarker profiles, and treatment history, to predict individual treatment response. By finding trends and connections in big datasets, AI systems can provide insights into the chance of treatment success, allowing healthcare workers to tailor treatments accordingly. b. Adverse Event Prediction: AI systems can study patient-specific data and known risk factors to predict the chance of adverse events or treatment-related problems. By finding patients at higher risk, healthcare workers can implement preventive steps or change treatment plans to reduce the occurrence of unfavorable events.

4. Integration of Real-World Data: a. Electronic Health Records (EHRs): AI can leverage EHR data to produce insights for treatment

plans and dose improvement. By studying a patient's medical background, test data, and treatment outcomes, AI systems can provide useful information for improving treatment plans. b. Real-World Evidence: AI models can examine real-world data, such as patient records, clinical studies, and post-marketing monitoring data, to create evidence for treatment plans and dose improvement. By considering data from different patient groups and real-world settings, AI can provide more complete and solid suggestions.

The inclusion of AI in treatment planning and dose optimization has the ability to increase patient results, improve treatment effectiveness, and reduce harmful events. However, issues such as data protection, algorithm openness, and legal factors need to be handled to ensure the responsible and ethical use of AI in clinical practice. Collaboration among healthcare workers, AI experts, regulatory bodies, and patient supporters is important for realizing the full potential of AI in treatment planning and dose optimization.

By tapping the power of AI, healthcare workers can optimize treatment methods, tailor doses to individual patients, and improve total treatment results, leading to more personalized and effective patient care.

3.4.2 Decision Support Systems for Treatment Recommendations

Decision support systems (DSS) driven by AI have emerged as useful tools in giving treatment advice for healthcare workers. These systems harness AI tools to evaluate patient data, clinical standards, and medical books to help in treatment decision-making. Let's look into how AI-driven decision support systems are changing treatment suggestions.

1. Integration of Patient Data: a. Electronic Health Records (EHRs): AI programs can examine patient data saved in EHRs, including medical information, test results, imaging reports, and

treatment records. By combining and handling this data, decision support systems can provide a complete picture of the patient's health state, allowing more informed treatment suggestions. b. tailored Patient Profiles: AI-driven decision support systems can build tailored patient profiles by considering factors such as demographics, genetic information, habits, illnesses, and treatment reaction. These profiles help doctors adjust treatment suggestions based on individual patient traits.

2. Evidence-Based Treatment Recommendations: a. Clinical Guidelines and Medical Literature: AI systems can study and understand vast amounts of clinical guidelines and medical literature. By pulling important information, decision support systems can provide evidence-based treatment suggestions matched with the latest study and best practices. b. Treatment Outcomes Analysis: AI can study real-world treatment outcomes data, including patient reactions to different treatment methods, to find trends and relationships. This study can help decision support systems produce treatment suggestions based on the chance of success and patient-specific factors.

3. Treatment Protocol Optimization: a. Treatment Sequencing and Timing: AI-driven decision support systems can consider disease development, treatment history, and patient traits to suggest best treatment sequences and timing. By considering the possible benefits and risks of different treatment choices, these tools help healthcare workers in creating effective treatment plans. b. customized Risk Assessment: AI systems can measure individual patient risk factors, such as genetic predispositions, illnesses, and drug combinations, to provide customized risk assessments. This allows decision support systems to suggest treatments that minimize risks and improve patient safety.

4. Integration with Clinical Workflows: a. Real-Time Decision Support: Decision support systems can be linked into healthcare

processes to provide real-time treatment suggestions. This combination allows healthcare workers to receive decision support at the point of care, enabling informed treatment choices without breaking clinical processes. b. Alerts and notes: AI-driven decision support systems can create alerts and notes based on patient-specific factors and treatment standards. These alerts help doctors stick to treatment guidelines, track treatment progress, and ensure quick responses.

5. Continuous Learning and Improvement: a. Machine Learning and comments Loops: AI systems can continuously learn and improve by studying comments from doctors and patient results. This iterative process helps decision support systems to improve treatment suggestions over time, adding new data and responding to individual patient reactions.

While AI-driven decision support systems offer useful aid in treatment suggestions, it's crucial to keep a human-in-the-loop method. Healthcare workers should understand the suggestions created by these systems within the context of their clinical knowledge and patient-specific factors.

The inclusion of AI in decision support systems improves treatment suggestions by leveraging extensive patient data, evidence-based standards, and real-world results. By combining clinical knowledge with AI-powered insights, healthcare workers can make well-informed treatment decisions, improve patient results, and provide personalized care.

3.4.3 AI in Surgical Planning and Intervention

AI technologies are changing surgery planning and assistance by offering improved tools for preoperative analysis, operating direction, and postoperative assessment. Let's study the ways in which AI is changing surgery processes.

1. Preoperative Planning: a. Medical Imaging Analysis: AI programs can study medical imaging data, such as CT scans, MRI pictures, and 3D models, to help in surgery planning. These systems can easily divide anatomical structures, spot flaws, and create 3D images, giving doctors with useful insights for treatment plans. b. Virtual Surgical simulators: AI-powered simulators can build virtual models of patient anatomy based on medical image data. Surgeons can use these models to practice surgical processes, study different methods, and measure the possible results, improving their understanding of the surgical process.

2. Intraoperative Guidance: a. Surgical guidance Systems: AI-driven surgical guidance systems utilize real-time image and tracking technologies to help doctors during procedures. These systems provide exact location, direction, and viewing of surgical tools and internal structures, helping in correct surgical procedures. b. Robotics-Assisted Surgery: AI plays a crucial role in robotics-assisted surgery, where intelligent robots perform medical procedures under the direction of doctors. AI algorithms allow robots to study real-time data, change moves, and enhance surgery accuracy, leading to better results and reduced invasiveness.

3. Augmented Reality (AR) and Virtual Reality (VR): a. AR and VR technologies mixed with AI can provide doctors with realistic visualizations and live advice during surgeries. Surgeons can add virtual information onto the patient's body, such as tumor sites, blood veins, or important structures, improving their knowledge and accuracy during interventions. b. Training and Education: AI-powered AR and VR models can be used for surgery training, allowing doctors to practice difficult techniques in a virtual setting. These models allow doctors to develop skills, improve methods, and gain experience in a risk-free setting.

4. Postoperative Assessment: a. automatic Postoperative Analysis: AI systems can study postoperative data, such as surgery

pictures, lab reports, and patient results, to provide automatic reviews. These methods can spot problems, measure surgery success, and find areas for change, helping in postpartum evaluation. b. Predictive Analytics: AI models can study postoperative data and patient results to create predictive analytics for surgery treatments. By finding trends and connections, these models can help doctors in predicting surgical problems, treatment reaction, and long-term results.

5. Data Integration and Collaboration: a. blending of Multimodal Data: AI technologies allow the blending of various data sources, such as medical images, genetics, and patient records. By mixing these data sets, doctors can access complete patient information and make informed choices based on an overall view of the patient's state. b. joint Platforms: AI-powered joint platforms help doctors to share information, skills, and surgery experiences. These sites allow the sharing of best practices, enable online counseling, and create a global community of medical experts.

AI-driven surgery planning and assistance equip doctors with advanced tools, real-time advice, and prediction powers, improving surgical accuracy, patient safety, and total results. However, it's important to ensure the responsible adoption of AI in surgery, handling concerns such as data protection, algorithm openness, and the need for medical skill in understanding AI-generated insights.

By adding AI technologies into surgical processes, healthcare workers can optimize surgery planning, increase clinical accuracy, and improve patient results, bringing in a new age of transformative surgical care.

3.5 Real-Time Monitoring and Adaptive Treatment
3.5.1 AI-Enabled Remote Patient Monitoring

Remote patient monitoring (RPM) refers to the use of technology to track patients' health state and vital signs outside of regular healthcare situations. With the merging of AI, online patient tracking has become more complex, allowing real-time monitoring, early discovery of changes in health states, and adaptable treatment methods. Let's study the role of AI in online patient tracking.

1. Continuous Data Collection: a. Wearable Devices: AI-enabled wearable devices, such as smartwatches, fitness trackers, and biosensors, can collect various bodily data, including heart rate, blood pressure, oxygen saturation, and exercise levels. These gadgets constantly watch the patient's health factors and send the data to healthcare workers or monitoring systems. b. Home tracking Systems: AI can combine with home tracking systems, allowing the collection of extra data, such as weight, blood levels, and drug adherence. AI systems examine this data to provide insights into the patient's health state and trends.

2. Real-Time Monitoring and Alerting: a. Data Analysis and Pattern Recognition: AI systems study the collected data in real-time, finding patterns and abnormalities that may suggest changes in the patient's health state. AI can spot trends, predict deteriorations, and create alerts for healthcare workers to act quickly. b. Predictive Analytics: By studying past data from multiple patients, AI models can predict the chance of bad events, hospital readmissions, or disease exacerbations. These models allow early treatments and effective control, lowering the chance of problems and improving patient results.

3. Personalized Health Insights: a. Machine Learning and Individualized Algorithms: AI algorithms can learn from individual patient data, building individualized models that change to the patient's unique health trends. This personalized method allows for tailored tracking and early discovery of changes from the patient's norm, improving the accuracy of alerts and actions. b. Behavioral Analysis and Risk Stratification: AI can study behavioral trends and risk factors to find people at higher risk of health deterioration. By

stratifying patients based on their risk factors, healthcare workers can allocate resources efficiently and focus on patients who require instant attention.

4. Adaptive Treatment Strategies: a. Intelligent Treatment suggestions: AI systems can combine remote tracking data with patient health information, treatment standards, and medical books to provide personalized treatment suggestions. These suggestions consider the patient's real-time health data, past treatment reactions, and individual traits, allowing for adaptable treatment methods. b. Virtual Care and Telemedicine: AI-powered remote patient tracking enables virtual care and telemedicine talks. Healthcare workers can directly watch patients, measure their situations, and change treatment plans without needing in-person trips. This method improves mobility, lowers healthcare costs, and boosts customer comfort.

5. Data Security and Privacy: a. Secure Data transfer: AI-enabled remote patient tracking systems apply secure data transfer methods to protect patient information. Encryption methods and strict data protection measures guarantee that patient data remains private during transfer and keeping. b. Compliance with Regulatory Standards: Remote patient tracking systems meet with important regulatory standards and rules to ensure patient data safety. These systems conform to data protection laws, such as HIPAA (Health Insurance Portability and Accountability Act), to keep the safety and security of patient information.

AI-enabled online patient tracking provides healthcare workers with real-time insights, early spotting of health changes, and flexible treatment methods. By adding AI technologies into remote tracking systems, healthcare workers can carefully control patient health, avoid problems, and improve total patient care.

3.5.2 Closed-Loop Systems for Adaptive Treatment

Closed-loop systems, also known as closed-loop control or feedback systems, are AI-enabled methods that constantly watch patient health and automatically change treatment measures based on real-time data. These systems create a feedback loop where data from tracking devices guide treatment choices, allowing flexible and personalized actions. Let's study the role of closed-loop devices in tailored medicine.

1. Continuous Monitoring and Data Collection:

• Closed-loop systems utilize various tracking technologies, such as wearable devices, sensors, or implantable devices, to collect real-time patient data. These devices record bodily factors, biomarkers, or other important health signs, giving a constant stream of data.

2. Data Analysis and Decision-Making:

• AI systems evaluate the collected data in real-time, finding patterns, trends, and oddities. These programs use machine learning and statistical techniques to analyze the data and produce useful insights.

• Decision-making models within closed-loop systems can be based on pre-defined rules, clinical guidelines, or machine learning algorithms that learn from past data to make treatment suggestions.

3. Treatment Intervention:

• Based on the study of patient data, closed-loop devices can automatically change medical interventions. These interventions may include drug amounts, drip rates, insulin dosing, or other treatment interventions.

• The closed-loop system constantly assesses the patient's reaction to treatment, analyzes the efficiency, and changes the solution in real-time to improve results.

4. Feedback Loop:

- Closed-loop systems create a feedback loop by constantly tracking the patient's bodily factors and changing treatment measures based on the feedback received.

- The system measures the impact of treatment changes and tracks the patient's reaction, giving ongoing feedback to improve treatment success.

5. Benefits of Closed-Loop Systems:

- customized Treatment: Closed-loop methods allow for customized treatment approaches tailored to the individual patient's needs and real-time health state.

- Real-Time Adaptation: With closed-loop devices, treatment changes can be made quickly based on real-time data, allowing for rapid solutions and improvement of patient care.

- Improved results: Closed-loop methods can help achieve better treatment results by ensuring that approaches are constantly changed and improved based on patient reaction.

- Reduced Burden on Healthcare Providers: By making treatment changes, closed-loop systems ease healthcare providers from the constant need to watch and physically adjust interventions, allowing them to focus on other critical aspects of patient care.

6. Challenges and Considerations:

- Safety and dependability: Closed-loop systems must face thorough testing and evaluation to ensure their safety and dependability in real-world situations.

- Data Accuracy and Quality: Accurate and high-quality data collection is important for closed-loop systems to make informed treatment choices.

• Ethical and Legal concerns: The use of closed-loop devices brings ethical and legal concerns, including patient liberty, informed permission, and responsibility problems.

• Regulatory Compliance: Closed-loop systems must meet with regulatory standards and rules to ensure patient safety and data privacy.

Closed-loop systems offer a hopeful method to tailored treatment by leveraging AI and real-time data analysis. These systems have the potential to increase treatment accuracy, improve patient results, and reduce the load on healthcare workers, eventually improving the field of personalized medicine.

3.5.3 AI in Real-Time Treatment Adjustment

Real-time treatment adjustment refers to the dynamic change of treatment measures based on constant monitoring and study of patient data. AI plays a crucial part in allowing real-time treatment change by handling large amounts of data, finding trends, and providing useful insights. Let's look into the application of AI in real-time treatment change.

1. Continuous Monitoring and Data Collection:

• AI-powered systems combine with tracking devices, wearable sensors, or internal devices to collect real-time patient data. These devices record vital signs, bodily factors, drug adherence, or other important data points.

• Data is passed to AI systems, which process and study the information in real-time.

2. Real-Time Data Analysis:

• AI systems evaluate the collected data, applying machine learning and pattern recognition techniques to extract useful insights.

- Real-time data analysis allows the discovery of minor changes, differences from normal ranges, or trends that suggest treatment reaction or possible bad events.

3. Treatment Optimization:

- Based on the study of real-time data, AI programs provide suggestions or make automatic changes to medical measures.

- Treatment changes can include drug doses, injection rates, time of treatments, or other therapeutic modifications.

- AI systems consider individual patient traits, past treatment reaction, and real-time health state to improve treatment efficiency and patient results.

4. Predictive Analytics:

- AI systems can predict future treatment reaction based on past data, treatment trends, and patient-specific factors.

- Predictive analytics help in forecasting possible treatment results, finding patients at risk of problems, or offering alternative treatment methods.

5. Feedback Loop and Learning:

- Real-time treatment change includes building a feedback loop between the AI system, patient data, and treatment actions.

- The AI system constantly watches the patient's reaction to treatment changes and adapts future actions based on feedback.

- The system learns from the results of previous changes and refines its methods to improve treatment optimization over time.

6. Clinical Decision Support:

- AI-powered clinical decision support tools provide real-time treatment suggestions to healthcare workers based on the study of patient data.

- These systems combine patient-specific data, medical books, treatment standards, and AI tools to support informed decision-making.

7. Benefits of AI in Real-Time Treatment Adjustment:

- Personalized Treatment: AI allows treatment changes suited to each patient's unique traits and real-time health state.

- Timely Interventions: Real-time data analysis allows for quick identification of treatment reaction or bad events, enabling timely interventions.

- Enhanced Treatment Precision: AI systems consider a multitude of factors and trends to improve treatment success and reduce side effects.

- Improved Patient results: Real-time treatment change based on AI research has the ability to improve patient results, treatment reaction rates, and general quality of care.

8. Challenges and Considerations:

- Data Quality and Integration: Ensuring accurate and reliable data collection and integration from various sources is important for effective real-time treatment change.

- ethics and Privacy Considerations: The use of patient data for real-time treatment change needs attention to privacy, permission, and ethics standards.

- Validation and Safety: Rigorous testing, validation, and legal compliance are important to ensure the safety and trustworthiness of AI systems for real-time treatment change.

AI Revolution in Medicine, The Future of Healthcare

AI-driven real-time treatment change holds promise in optimizing patient care, allowing personalized actions, and better treatment results. As technology improves and AI programs continue to change, the potential for real-time treatment adjustment to transform healthcare service becomes increasingly important.

3.6 Ethical Considerations in AI-Driven Treatment

3.6.1 Data Privacy and Security

The use of AI-driven treatment in healthcare raises important social considerations, especially regarding data safety and security. As AI systems rely on vast amounts of patient data for research and decision-making, it is important to address the following problems linked to data protection and security:

1. told Consent: Patients should be fully told about the gathering, keeping, and use of their data for AI-driven care. Obtaining informed permission ensures that people understand the effects of sharing their data and have the right to control how their information is utilized.

2. Data security: Healthcare workers and organizations must adopt strong means to protect patient data and ensure its security. This includes safe storage, encryption, access controls, and tight obedience to data security laws such as HIPAA (Health Insurance Portability and Accountability Act) in the United States.

3. Anonymization and De-identification: To minimize privacy risks, patient data used for AI-driven care should be properly masked or de-identified. This process removes personally identifying information, lowering the chance of re-identification and protecting patient privacy.

4. Data Access and Sharing: AI programs may require access to diverse datasets to enhance their accuracy and generalizability. However, the sharing of patient data should only be done with

proper permission and in line with privacy rules to ensure the protection of patient privacy.

5. Security Measures: Robust security measures, including safe networks, encryption protocols, and identification methods, should be adopted to prevent illegal access, data breaches, and cyber-attacks. AI systems must face thorough testing and evaluation to find and solve possible flaws.

6. Transparent Data Governance: Clear rules and policies should be set to control the collection, keeping, and use of patient data for AI-driven care. Transparency in data control builds trust among patients and ensures that their data is treated properly.

7. Ethical Use of Data: Healthcare groups and AI makers must stick to ethical standards and principles when utilizing patient data. This includes using the data solely for the intended purposes, avoiding biased practices, and favoring patient well-being.

8. Regulatory Compliance: Compliance with relevant data protection and privacy laws, such as the General Data Protection Regulation (GDPR) in the European Union, is important. Adhering to these rules guarantees that patient data is treated in line with legal requirements, protecting individual rights and privacy.

9. Data keeping and removal: Clear rules should be in place for the keeping and removal of patient data once it is no longer needed. This ensures that data is not kept forever and is safely disposed of to prevent any illegal access or misuse.

10. Continuous Monitoring and Auditing: Regular monitoring and auditing of AI systems and data practices help spot and solve any possible privacy or security issues. Ongoing review and improvement of data protection methods are important to keep a high level of security.

AI Revolution in Medicine, The Future of Healthcare

By handling data protection and security issues, healthcare organizations can promote trust among patients and ensure the responsible use of AI-driven treatment. Ethical factors should lead the development, adoption, and continuing management of AI systems to protect patient privacy and keep the secrecy and integrity of their data.

3.6.2 Bias and Fairness in Treatment Recommendations

The application of AI in treatment suggestions brings the need to handle bias and ensure fairness in the decision-making process. AI systems can accidentally spread biases present in the data used for training, leading to possible inequalities in treatment suggestions. Here are some factors to promote fairness and minimize bias in AI-driven treatment recommendations:

1. Data Representation: Ensure that the training data used to build AI systems is representative of different patient groups. Adequate representation helps avoid underrepresentation or misunderstanding of certain ethnic groups, lowering the risk of biased results.

2. Bias Detection and Mitigation: Implement methods to identify and reduce bias in AI systems. Regularly review the success of the system across different classes and measure if there are any differences in treatment suggestions. If errors are discovered, take proper steps to understand their causes and change the program accordingly.

3. Transparent Algorithmic Processes: Foster openness in AI systems by giving reasons for treatment suggestions. Users, including healthcare workers and customers, should be able to understand the factors that lead to the suggestions. Transparent algorithms allow for examination and discovery of possible flaws.

4. Diverse Development Teams: Encourage variety within the teams responsible for building AI programs and systems. Including

people with various backgrounds and views helps discover and solve biases that might otherwise go unseen.

5. Regular Evaluation and Validation: Continuously watch and evaluate the performance of AI systems to find and correct any flaws that appear over time. This ongoing review ensures that the method stays fair and neutral as new data becomes available.

6. Ethical Guidelines and Standards: Develop and stick to ethical guidelines and standards specific to AI-driven treatment suggestions. These standards should clearly address problems of bias, fairness, and justice, leading the development and application of AI systems in healthcare.

7. Patient Empowerment and Shared Decision-Making: Empower patients to actively join in the decision-making process by giving them with clear accounts of treatment guidelines. Encourage shared decision-making, where patients can talk their opinions, values, and worries with healthcare workers.

8. Regular Auditing and review: Establish independent auditing and review methods to ensure responsibility and fairness in AI-driven treatment suggestions. These methods can help spot and fix any flaws that may result from the use of AI systems.

9. Addressing Data Bias: Pay attention to possible flaws present in the training data and take steps to address them. This may involve using debiasing methods, carefully organizing various datasets, or using techniques such as counterfactual fairness to ensure fairness in treatment suggestions.

10. Continuous Education and Awareness: Educate healthcare workers, academics, and stakeholders about the possible flaws and challenges connected with AI-driven treatment suggestions. Promote understanding of the value of justice and equality in healthcare decision-making processes.

AI Revolution in Medicine, The Future of Healthcare

Addressing bias and ensuring fairness in AI-driven treatment suggestions is important to provide equal healthcare results for all patients. By adopting these factors, healthcare groups can work towards reducing bias and supporting fair and neutral treatment options.

3.6.3 Informed Consent and Transparency

Informed permission and openness are important ethical factors when deploying AI in treatment. Patients must have a clear understanding of the effects, risks, and benefits involved with AI-driven treatment suggestions. Here are key things to consider regarding informed agreement and transparency:

1. Clear Communication: Healthcare workers and AI writers should speak in clear and understandable English, avoiding technical words, when addressing AI-driven treatment advice with patients. They should explain how AI is being used, what data is being collected, how it affects treatment choices, and any possible limits or doubts.

2. Education and information: Promote patient education and information regarding the use of AI in medicine. This can include giving information about the powers, limits, and possible results of AI-driven care, as well as tools for patients to seek additional information or support.

3. Shared Decision-Making: Encourage shared decision-making between healthcare workers and customers. Provide patients with the chance to share their tastes, values, and worries, and involve them in the decision-making process regarding the use of AI-driven treatment suggestions.

4. permission Forms: Develop special permission forms that describe the use of AI in care and clearly seek patient approval for its application. These forms should clearly explain the goal of AI,

the possible risks and benefits, data usage, privacy factors, and the patient's rights regarding their data.

5. openness in Algorithms: Strive for openness in AI algorithms used for treatment suggestions. While protecting private information, provide healthcare workers and patients with insights into how the algorithms work, including the factors considered, the weightage given to different variables, and any possible flaws that may appear.

6. Explanation of suggestions: Ensure that AI-driven treatment suggestions can be explained to patients in an open and understandable way. Patients should have access to details regarding the reasoning behind the suggestions, the evidence backing them, and any doubts or limits.

7. Privacy Protection: Safeguard patient privacy and security throughout the AI-driven treatment process. Clearly explain the steps taken to protect patient data, including encryption, safe keeping, and obedience to important privacy laws.

8. Continuous Monitoring and Evaluation: Continuously watch and evaluate the success of AI systems used in treatment suggestions. This includes regularly analyzing their correctness, usefulness, and possible flaws. Patients should be told about these reviews and any updates or changes to the formulas that may affect their care.

9. Patient Rights and Control: Respect and protect patient rights regarding their data and treatment decisions. Provide patients with the ability to view their data, understand how it is being used, and have the option to remove consent for AI-driven treatment suggestions if wanted.

10. Patient Support and Redress: Establish methods for patient support and redress in situations where worries or issues arise

regarding AI-driven treatment suggestions. Provide ways for patients to seek explanation, share worries, and request further information or different treatment choices.

By valuing informed consent and openness, healthcare workers and AI writers can ensure that patients are actively involved in the treatment decision-making process. Transparent communication and patient initiative help build trust, support autonomy, and defend the ethical ideals of beneficence and care for patient autonomy.

3.7 Regulatory and Adoption Challenges in AI Treatment

3.7.1 Regulatory Framework for AI in Treatment

The introduction of AI in treatment presents several legal hurdles due to the unique nature of these technologies and their possible impact on patient care. Establishing a strong legal structure is important to ensure patient safety, ethical use of AI, and successful merging of these technologies into clinical practice. Here are key factors for building a legal system for AI in treatment:

1. Regulatory Authorities: Designate regulatory authorities responsible for overseeing AI systems in healthcare. These officials should have experience in AI, medicine, and legal processes to successfully review and watch AI-based treatments.

2. Risk-Based Approach: Implement a risk-based approach to regulatory control, defining AI apps based on their possible risks to patient safety and the complexity of the technology. This method helps assign regulatory resources more effectively, focusing attention on high-risk apps.

3. Data Privacy and Security: Develop laws to ensure the privacy and security of patient data used in AI-driven therapy. These rules should match with current data security laws and stress the responsible use and keeping of medical data.

4. Pre-Market Assessment: Establish a thorough pre-market assessment method for AI-based medical solutions. This method should evaluate the safety, effectiveness, and performance of the technology, including factors of bias, openness, and clinical proof.

5. Clinical Validation and Evidence: Define standards for clinical validation and evidence creation to show the success and safety of AI-driven medical tools. This may involve performing well-designed clinical trials, real-world data collection, or comparing efficiency studies to measure the impact on patient results.

6. Post-Market Surveillance: Implement post-market surveillance methods to watch the safety and success of AI-based medicines after they hit the market. These methods should allow continued data collection, bad event reporting, and review of real-world success.

7. Interoperability and Standardization: Encourage interoperability and standardization of AI technologies in treatment to enable smooth integration into current healthcare systems. This includes creating compatible data forms, application programming interfaces (APIs), and guidelines for data sharing.

8. Ethical Guidelines: Incorporate ethical guidelines into the legal system to address the ethical concerns related with AI-driven solutions. These recommendations should cover problems such as data protection, openness, bias prevention, patient liberty, and fair access to AI technologies.

9. International Collaboration: Foster international collaboration and harmonization of legal standards to enable the global usage and trade of AI-driven therapy technologies. Collaboration among regulatory bodies can promote clarity, simplify regulatory processes, and ensure patient safety across countries.

AI Revolution in Medicine, The Future of Healthcare

10. constant Evaluation and Adaptation: Establish a framework for constant evaluation and adaptation of the regulatory structure as technology improves and new problems appear. Regularly measure the success of the laws, solicit comments from stakeholders, and make necessary changes to meet new problems.

Developing a complete legal framework for AI in treatment needs teamwork among policymakers, healthcare workers, business players, and patient support groups. The structure should strike a mix between encouraging innovation and ensuring patient safety, privacy, and ethical use of AI tools in healthcare.

3.7.2 Integration of AI in Clinical Practice

The integration of AI in clinical practice offers various challenges and factors that need to be handled for successful acceptance and utilization. Here are key factors to consider for the successful application of AI in clinical practice:

1. Education and Training: Provide thorough education and training programs for healthcare workers to build the necessary skills and information to successfully use AI technologies. This includes training on data analysis, understanding AI systems, and adding AI-driven insights into healthcare decision-making.

2. Workflow Integration: Adapt current healthcare workflows to add AI technologies smoothly. Identify areas where AI can improve speed and accuracy, and add AI tools and apps into the healthcare routine to simplify processes. This may involve rethinking processes, combining AI software with current electronic health record systems, and ensuring interoperability with other healthcare tools.

3. Human-AI Collaboration: Foster a joint method between healthcare workers and AI platforms. Emphasize the role of AI as a supporting tool rather than a substitute for human knowledge.

Encourage healthcare workers to actively interact with AI tools, understand AI-generated insights, and make informed choices based on their clinical opinion.

4. Evidence-Based Practice: Emphasize the value of evidence-based practice when applying AI in healthcare decision-making. AI algorithms should be tested using solid clinical data, and their performance should be reviewed in real-world settings to ensure their accuracy, dependability, and clinical usefulness.

5. legal Compliance: Ensure that the inclusion of AI technologies in clinical practice adheres to legal rules and standards. Healthcare companies should follow with data protection rules, ethics standards, and any special laws linked to the use of AI in healthcare. Regular checks and reviews can help ensure compliance.

6. Data Management and Governance: Develop effective data management and governance methods to handle the large amounts of data gathered by AI systems. This includes ensuring data quality, handling data biases, adopting suitable data security measures, and creating processes for data sharing and keeping.

7. Clinical Validation and Evaluation: Prioritize the clinical validation and evaluation of AI technologies before their introduction into regular clinical practice. Rigorous validation studies should measure the performance, accuracy, safety, and clinical effect of AI technologies in important patient groups. Continuous tracking and assessment should also be performed post-implementation to measure long-term success and solve any new problems.

8. Ethical factors: Incorporate ethical factors into the merging of AI in clinical practice. This includes handling problems related to privacy, openness, fairness, bias reduction, informed consent, and equal access to AI technologies. Ethical principles and models can

help guide decision-making and ensure the responsible and ethical use of AI in healthcare.

9. Patient-Centered method: Maintain a patient-centered method throughout the merging of AI in healthcare practice. Focus on patient involvement, teaching, and shared decision-making to ensure that patients are well-informed and fully involved in their care. AI technologies should be utilized to improve patient results, increase patient experiences, and enable patients to make informed healthcare choices.

10. Continuous Learning and Improvement: Embrace a mindset of continuous learning and improvement when applying AI in healthcare practice. Regularly examine the effect of AI technologies, receive feedback from healthcare experts and patients, and make necessary changes to improve the merging process. Emphasize teamwork, open communication, and a desire to change and improve practices based on real-world experiences.

By considering these factors, healthcare companies can successfully combine AI technologies into clinical practice, leading to better patient care, enhanced decision-making, and breakthroughs in individualized medicine.

3.7.3 Addressing Barriers to Adoption of AI in Treatment

The adoption of AI in treatment holds great promise for better patient results and changing healthcare. However, several hurdles exist that prevent the broad integration of AI technologies into clinical practice. Addressing these hurdles is important to support the acceptance of AI in treatment. Here are key tactics to beat hurdles to adoption:

1. recognition and Education: Increase recognition and learning of AI among healthcare workers, lawmakers, and patients. Provide thorough education and training programs to familiarize

stakeholders with AI ideas, benefits, and limits. This includes classes, lectures, online training, and partnerships with university organizations.

2. proof-Based Demonstrations: Conduct thorough studies and demonstrations to provide clear proof of the usefulness and value of AI in treatment. Generate high-quality clinical proof through well-designed studies and real-world reviews. Demonstrating better patient results, greater speed, and cost-effectiveness can build trust in healthcare workers and decision-makers.

3. Interoperability and Integration: Promote interoperability and smooth integration of AI technologies with current healthcare systems and infrastructure. Develop uniform data forms, application programming interfaces (APIs), and communication frameworks to support the sharing of data between AI systems and healthcare tools. This guarantees that AI smoothly fits into the process and does not cause additional loads or disruptions.

4. Addressing Data issues: Address issues relating to data quantity, quality, and usability. Develop data-sharing programs, data control frameworks, and safe data-sharing tools to enable the sharing of anonymous patient data for AI training and evaluation. Ensure data protection and security while keeping compliance with relevant laws and ethics guidelines.

5. legal Framework: Establish clear legal rules and frameworks specific to AI in treatment. Regulators should provide guidance on the review, approval, and post-market monitoring of AI technologies. This includes handling problems such as data protection, openness, bias prevention, and patient safety. Collaboration between governing bodies, business partners, and healthcare workers is important in making suitable laws.

6. Ethical concerns: Prioritize ethical concerns in the development and use of AI in medicine. Address issues regarding

privacy, permission, openness, and fairness. Develop ethical rules and models that lead the responsible and fair use of AI technologies. Promote openness in AI systems and decision-making processes to build trust among healthcare workers and patients.

7. Cost Considerations: Address cost-related issues linked with AI usage. Conduct cost-effectiveness studies and economic reviews to show the long-term benefits and possible cost savings related with AI integration. Explore payment methods and funding mechanisms that incentivize the adoption of AI technologies while ensuring cost and usability for healthcare workers and customers.

8. Collaboration and Partnerships: Foster alliances and partnerships between healthcare groups, technology companies, education schools, and governmental bodies. Encourage information sharing, study partnerships, and innovation networks to leverage the skills of different parties. Collaborative efforts can speed the creation, approval, and uptake of AI tools in treatment.

9. User-Friendly Interfaces: Develop user-friendly interfaces and tools that are simple and easy to use for healthcare workers. AI technologies should easily blend into clinical processes and provide useful insights without adding complexity or stress to the healthcare team. customer-centered design concepts can improve customer acceptance and usage.

10. ongoing Evaluation and Improvement: Establish methods for ongoing evaluation, feedback, and improvement of AI systems in treatment. Monitor the success, safety, and effect of AI systems in real-world settings. Gather feedback from healthcare experts and patients to find areas for improvement and refine the technology accordingly.

By adopting these tactics, healthcare groups can overcome roadblocks to the acceptance of AI in treatment. This will enable the merging of AI technologies into regular clinical practice, eventually leading to better patient results, enhanced speed, and breakthroughs in personalized medicine.

3.8 Future Perspectives and Impact of AI in Personalized Medicine

3.8.1 AI and the Future of Targeted Therapies

AI has the ability to change the research and delivery of focused treatments in personalized medicine. By leveraging large-scale data analysis, AI systems can find molecular fingerprints, biomarkers, and genetic differences linked with specific illnesses. This allows the finding of new drug targets and the creation of customized treatments that meet the unique features of individual patients.

AI programs can examine complicated biological data, such as genetics, proteomics, and metabolomics, to find possible drug targets and predict the effectiveness of focused treatments. This allows researchers and doctors to pick the most suitable treatment choices for patients based on their specific genetic patterns, raising the chance of treatment success and reducing harmful effects.

Furthermore, AI can simplify the design of clinical studies for tailored treatments by finding patient groups most likely to gain from the treatment. AI programs can examine diverse patient data, including personal information, genetic profiles, and disease traits, to find classes of patients who are more responsive to specific treatments. This improves the speed of clinical studies and increases the chance of good results.

3.8.2 AI-Driven Clinical Trials and Evidence Generation

AI has the ability to change the world of clinical studies and proof creation in personalized medicine. By leveraging AI technologies, researchers can optimize study design, find suitable patient groups, and improve patient recruitment and retention.

AI programs can examine big amounts of patient data from electronic health records, DNA databases, and clinical trial databases to find possible candidates for specific clinical studies. This speeds patient recruitment, improves study effectiveness, and lowers costs.

Moreover, AI can support real-time tracking and analysis of study data, allowing researchers to identify treatment reactions, bad events, and patient results in a more fast and accurate way. This allows adaptable trial designs, where treatment methods can be changed based on ongoing data analysis, leading to more efficient and effective studies.

Additionally, AI can add to evidence creation by studying and combining data from multiple sources, including clinical studies, real-world data, and patient-reported results. This allows the creation of solid data on the safety, efficiency, and cost-effectiveness of tailored treatments in various patient groups.

3.8.3 The Role of AI in Patient Empowerment and Shared Decision-Making

AI technologies play a crucial part in enabling patients and promoting shared decision-making in personalized medicine. By giving patients with access to their health data, AI-powered tools allow patients to actively join in their own care and make informed choices.

AI systems can study patient-specific data, including genetic information, treatment history, and clinical results, to provide personalized treatment suggestions and predictive insights. This enables patients by providing them with complete information about their situation, possible treatment choices, and predicted results, enabling shared decision-making between patients and healthcare workers.

Furthermore, AI-driven platforms can provide teaching materials, risk assessment tools, and decision support systems that allow patients to understand the risks and benefits of different treatment choices. Patients can participate in conversations with healthcare workers, ask relevant questions, and actively add to the development of individual treatment plans.

AI technologies also have the ability to allow online tracking and self-management of chronic illnesses. Wearable devices and mobile apps driven by AI algorithms can track patient vital signs, provide personalized health suggestions, and inform patients and healthcare workers about any changes from normal health parameters. This allows early action, lowers hospital trips, and improves total patient results.

In summary, AI technologies have a major effect on the future of personalized health. They allow the development of tailored treatments, improve clinical studies, empower patients, and increase shared decision-making. The merging of AI in personalized medicine holds great potential for better patient results, cutting healthcare costs, and advancing the field of healthcare.

CHAPTER 4: AI IN MEDICAL IMAGING

4.1 Introduction to Medical Imaging

Medical imaging plays a crucial part in modern healthcare by offering useful views into the human body's internal structures and functions. It includes the use of different imaging technologies to view and identify diseases, guide treatment decisions, and track patient progress. This part explains the idea of medical imaging and shows its importance in healthcare.

4.1.1 Importance of Medical Imaging in Healthcare

Medical imaging serves as a powerful tool for healthcare workers in several ways:

1. Diagnosis and Disease Detection: Medical imaging methods such as X-rays, computed tomography (CT), magnetic resonance imaging (MRI), and ultrasound allow the viewing of physical structures and the identification of abnormalities or illnesses. They help in identifying situations like fractures, cancer, infections, circulatory diseases, and brain problems.

2. Treatment Planning and Intervention: Medical imaging helps in planning and leading different medical treatments and actions. For example, imaging technologies like CT and MRI provide detailed pictures that help in surgery planning, radiation therapy targets, and image-guided actions. This provides accuracy and reduces risks during processes.

3. watching and Follow-up: Imaging methods play a vital role in watching disease development, treatment reaction, and post-treatment healing. Sequential imaging scans can help measure changes in tumor growth, spot spread, or evaluate the success of treatment approaches. Imaging results help healthcare workers in making suitable changes to treatment plans.

AI Revolution in Medicine, The Future of Healthcare

4. Early discovery and Prevention: Medical imaging allows for the early discovery of illnesses, allowing quick action and better patient results. Screening programs, such as mammograms for breast cancer or lung cancer screening with low-dose CT, can help identify problems at an early stage when they are more manageable.

5. Minimally Invasive operations: Advanced imaging technologies, such as fluoroscopy and ultrasound, allow real-time viewing during minimally invasive operations. They help guide the placement of tubes, needles, or surgical tools, lowering the need for open surgery and limiting patient stress and healing time.

6. study and Medical Education: Medical imaging adds to medical study by giving useful data for scientific studies, clinical trials, and the development of new imaging methods. It also plays a vital part in medical education by helping students and healthcare workers to understand and study physical structures and abnormal situations.

Overall, medical imaging plays a vital role in the detection, treatment, and control of different illnesses. It improves patient results, increases the accuracy of medical treatments, and adds to the development of medical understanding. The inclusion of artificial intelligence (AI) in medical imaging further improves its powers, allowing automatic analysis, better accuracy, and personalized healthcare methods.

4.1.2 Overview of Different Medical Imaging Modalities

Medical imaging includes a range of methods that apply different techniques to record pictures of the human body. Each mode has its unique traits, benefits, and uses. This part gives an outline of the widely used medical imaging modalities:

1. X-ray: X-ray imaging uses ionizing radiation to make pictures of the body's internal structures. It is frequently used to view bones

and identify injuries, lung problems, and certain stomach conditions. X-rays are quick, non-invasive, and widely available.

2. Computed Tomography (CT): CT scans use a number of X-ray pictures taken from different points to make accurate cross-sectional views of the body. CT offers more complete information than X-rays and is used for identifying various conditions, including cancer, arterial diseases, and accidents. It is particularly useful for viewing internal organs and complicated structural systems.

3. Magnetic Resonance Imaging (MRI): MRI uses strong magnetic fields and radio waves to create detailed pictures of the body's soft tissues and organs. It is particularly useful in viewing the brain, spinal cord, joints, and soft organs. MRI is useful for identifying brain diseases, musculoskeletal issues, and finding tumors.

4. Ultrasound: Ultrasound imaging uses high-frequency sound waves to make real-time pictures of the body's features. It is safe, non-invasive, and widely used in pediatrics, medicine, and heart. Ultrasound is useful for viewing organs, measuring blood flow, and directing invasive treatments.

5. Positron Emission Tomography (PET): PET imaging includes the use of a radioactive source that creates positrons. The scanner identifies these particles to make pictures showing biochemical and cellular processes in the body. PET is often used for cancer stages, analyzing brain diseases, and measuring heart performance.

6. Single-Photon Emission Computed Tomography (SPECT): SPECT uses nuclear tracers similar to PET, but it identifies the gamma-ray bursts instead of positrons. SPECT is widely used for measuring blood flow, brain diseases, and certain heart conditions.

7. Nuclear Medicine: Nuclear medicine involves the administration of nuclear materials (radiopharmaceuticals) to

identify and treat illnesses. Gamma cameras or SPECT/CT and PET/CT machines are used to identify the released radiation and make pictures showing the spread of the radiopharmaceuticals in the body.

These imaging technologies have changed medical evaluation and treatment, allowing healthcare workers to view internal structures, spot abnormalities, and guide treatments. The inclusion of AI technologies in medical imaging further improves the accuracy, speed, and understanding of these imaging methods, opening new opportunities for individualized medicine and better patient care.

4.2 Evolution of AI in Medical Imaging

Medical imaging has experienced a significant change with the introduction of artificial intelligence (AI) technologies. AI programs have shown great promise in improving the accuracy, speed, and understanding of medical imaging studies. This part discusses the development of AI in medical imaging and shows some early uses.

4.2.1 Early Applications of AI in Medical Imaging

The early uses of AI in medical imaging worked on building methods for automatic picture analysis and computer-aided detection. Here are some famous examples:

1. Image Segmentation: AI algorithms were created to divide medical images and find specific structures or areas of interest. This enabled the quantification and study of morphological traits and clinical abnormalities.

2. Lesion Detection and Classification: AI systems were taught to identify and describe lesions in medical pictures, such as cancer in CT or MRI scans. These systems helped doctors in finding possible problems and selecting cases for further evaluation.

3. Radiomics: Radiomics involves the extraction of numeric traits from medical pictures, which can be used to predict disease results, treatment reaction, or risk assessment. AI systems played a crucial part in studying big collections of radiomic traits and finding trends related with specific diseases.

4. Computer-Aided Detection (CAD): CAD systems were created to help doctors in spotting suspicious finds in medical pictures. These methods flagged possible abnormalities, such as lumps or microcalcifications, to help in the early discovery of diseases like breast cancer or lung cysts.

5. Image Reconstruction and Enhancement: AI methods were applied to improve image quality, reduce noise, and increase the visualization of features in medical photos. This led to better and more accurate pictures, improving diagnosis capabilities.

6. Image Registration: AI systems allowed the registration of multiple imaging methods or photos taken at different time points. This allowed the comparison and fusion of pictures for better viewing and measurement of changes over time.

These early applications of AI in medical imaging set the basis for more advanced and complex uses of AI in the field. The merging of AI technologies in medical imaging holds great promise for improving diagnosis accuracy, cutting analysis time, and allowing personalized treatment strategies. As AI continues to grow, its possible effect on medical imaging is likely to expand, leading to further breakthroughs in patient care and results.

4.2.2 Technological Advances Enabling AI in Medical Imaging

The fast progress in technology has played a key role in allowing the application of artificial intelligence (AI) in medical imaging. Several technical advances have added to the merging of AI systems into medical imaging processes. Here are some key technological breakthroughs that have allowed the acceptance of AI in medical imaging:

1. Increased computer Power: The availability of high-performance computer facilities, including graphics processing units (GPUs) and cloud computing, has greatly enhanced the working powers needed for AI programs. These strong computer tools allow the fast training and performance of complex AI models.

2. Big Data and Digitalization: The digitalization of medical imaging data and the spread of electronic health records (EHRs) have created vast amounts of imaging data for analysis. AI systems rely on big data, as they require large datasets for training and evaluation. The availability of digital image files and defined data forms has made it easier to access and utilize these datasets.

3. Deep Learning and Neural Networks: Deep learning, a type of machine learning, has changed AI in medical imaging. Deep neural networks, made of multiple levels of linked artificial neurons, shine at understanding complex patterns and traits from medical pictures. This has led to amazing advances in picture recognition, segmentation, and classification jobs.

4. Image Preprocessing and Augmentation: Preprocessing methods, such as noise reduction, image registration, and normalization, have improved the quality and uniformity of medical photos. Augmentation methods, such as data augmentation and generative adversarial networks (GANs), have increased the available training data and improved the generalizability of AI models.

5. High-Resolution Imaging: Technological improvements in medical imaging devices have led to better spatial resolution and improved picture quality. High-resolution imaging allows better viewing of internal structures and clinical abnormalities, giving greater data for AI systems to study.

AI Revolution in Medicine, The Future of Healthcare

6. Integration of Clinical Data: AI systems can leverage not only medical imaging data but also other important clinical data, such as patient details, clinical history, and genomic information. Integrating these diverse data sources improves the context and accuracy of AI-driven studies and predictions.

7. Data Sharing and Collaborative Platforms: The growth of data-sharing programs and collaborative platforms has enabled the pooling of medical imaging data from multiple organizations and study networks. This joint effort has allowed the creation of bigger and more diverse datasets for teaching AI algorithms, leading to more stable and flexible models.

These technology developments have opened the way for the successful integration of AI in medical images. They have allowed the development of advanced AI systems capable of studying complicated imaging data, improving diagnosis accuracy, and helping doctors in making more informed choices. As technology continues to progress, the possibility for AI in medical images will continue to grow, further changing patient care and results.

4.3 AI in Nuclear Medicine and Molecular Imaging

Nuclear medicine and molecular imaging play an important part in identifying and tracking different illnesses at the molecular and cellular levels. Artificial intelligence (AI) has emerged as a powerful tool in these areas, allowing the quantitative study and understanding of molecular imaging data. This area focuses on the application of AI in nuclear medicine and molecular imaging, with a particular stress on the quantitative study of molecular imaging data.

4.3.1 Quantitative Analysis of Molecular Imaging Data

chemical imaging methods, such as positron emission tomography (PET) and single-photon emission computed tomography (SPECT), provide functional and chemical details about the human body. AI programs have been created to perform quantitative analysis of molecular image data, allowing for more exact and objective readings. Here are some key uses of AI in the mathematical study of molecular imaging data:

1. Image rebuilding: AI systems have been applied to improve the quality of molecular imaging data through advanced image rebuilding methods. These methods can improve picture sharpness, reduce noise, and increase overall image quality, leading to more accurate numeric analysis.

2. Lesion Segmentation: AI systems can automatically separate and outline tumors in molecular imaging pictures. By correctly defining the boundaries of tumors, these methods enable quantitative measures, such as volume, metabolic activity, and variability. This information is important for disease stage, treatment reaction rating, and prediction.

3. Radiomics Analysis: Radiomics includes the gathering and analysis of a large number of numeric traits from molecular imaging data. AI systems can quickly study radiomic traits and find image biomarkers associated with specific diseases. These factors can help in identifying treatment reaction, stratifying patient risk, and directing individual treatment choices.

4. Kinetic Modeling: AI programs can be used to describe the dynamics of radiotracers in molecular imaging studies. By studying the dynamic changes of tracer uptake and removal over time, these methods can provide useful insights into physiological processes, such as blood flow, metabolism, and receptor binding. Kinetic modeling helps in the measurement of functional factors, such as standardized uptake values (SUVs) and kinetic rate constants.

5. Image Registration and Fusion: AI methods allow the registration and fusion of molecular imaging data with other imaging techniques, such as computed tomography (CT) or

magnetic resonance imaging (MRI). This multimodal union improves tissue localization and allows the merging of genetic and structural information for more thorough analysis.

6. Disease Classification and Prediction: AI systems can learn from big collections of molecular imaging data to identify illnesses or predict patient results. By finding trends and relationships within image data, these programs can help in alternative diagnosis, treatment planning, and patient classification.

The application of AI in the mathematical study of molecular imaging data holds great promise for better diagnosis accuracy, treatment planning, and patient care in nuclear medicine. By leveraging AI algorithms, doctors and academics can unlock the full potential of molecular imaging data, leading to increased knowledge of diseases, more focused treatments, and better patient results.

4.3.2 Radiopharmaceutical Design and Optimization

Radiopharmaceuticals are important components of nuclear medicine monitoring and treatment. They consist of a radioactive isotope mixed with a medicine substance that targets particular organs or signals in the body. AI methods have been applied to the design and improvement of radiopharmaceuticals, improving their effectiveness and widening their uses. Here are some key features of AI in radiopharmaceutical creation and optimization:

1. Target Identification: AI systems can examine big collections of molecular and biological information to find possible targets for radiopharmaceuticals. By collecting genetic data, proteome data, and other related information, AI can find specific biomarkers or molecular processes associated with illnesses. This information helps in picking suitable targets for radiopharmaceutical creation.

2. Virtual Screening: AI methods, such as machine learning and molecule docking, can be applied for virtual screening of possible

radiopharmaceutical chemicals. These programs examine chemical shapes and qualities to predict the binding affinity and specificity of potential drugs to the target of interest. Virtual screening speeds the finding of lead compounds and lowers the need for thorough laboratory screening.

3. Pharmacokinetics Modeling: AI systems can model the pharmacokinetics of radiopharmaceuticals, predicting their behavior in the body after treatment. By considering factors such as absorption, diffusion, metabolism, and elimination, these models help in improving dose, image collection time, and treatment planning. This ensures that the radiopharmaceutical gives the desired diagnostic or treatment benefit while minimizing harmful effects.

4. Image study and measurement: AI methods allow the study and measurement of radiopharmaceutical uptake in imaging studies. These programs can easily divide areas of interest, extract numeric data, and create standardized uptake values (SUVs) or other relevant metrics. This quantitative study aids in correct identification, treatment reaction assessment, and tracking disease development.

5. Dosimetry Optimization: AI systems can improve the dosimetry of radiopharmaceutical treatment. By considering patient-specific factors, such as organ shape, function, and radiation sensitivity, these programs can figure the best dose distribution for focused treatment while reducing radiation exposure to healthy tissues. Dosimetry tuning improves the effectiveness and safety of radiopharmaceutical medicines.

6. Theranostics: AI methods play a vital part in theranostics, which includes the merging of diagnostic and treatment functions in a single radiopharmaceutical. AI programs can help in the design and development of theranostic agents by finding ideal mixtures of

diagnostic and therapeutic isotopes, improving doses, and predicting treatment results based on patient traits.

The application of AI in radiopharmaceutical creation and optimization holds significant promise for better personalized medicine methods in nuclear medicine. By leveraging AI algorithms, researchers and doctors can enhance the precision, sensitivity, and therapeutic effectiveness of radiopharmaceuticals, leading to more accurate diagnoses, focused treatments, and better patient results.

CHAPTER 5: AI AND ELECTRONIC HEALTH RECORDS

5.1 Introduction to Electronic Health Records (EHR)
5.1.1 Definition and Components of EHR

In this part, we will study the concept of Electronic Health Records (EHR) and explain the major components that make up an EHR system.

Electronic Health Records (EHR) refer to digital representations of a patient's medical information and health history. An EHR is meant to acquire, preserve, and communicate patient data in a safe and standardized electronic manner. It serves as a complete record of a patient's health information, including medical history, diagnosis, prescriptions, test results, imaging reports, and treatment plans. Unlike conventional paper-based records, EHRs provide significant benefits in terms of accessibility, efficiency, and data interchange.

The components of an EHR system generally include:

1. Demographic Information: This covers the patient's personal facts such as name, age, gender, contact information, and insurance information. It offers a basic profile of the patient.

2. Medical History: The EHR includes the patient's medical history, including prior illnesses, surgeries, allergies, and family medical history. This information helps healthcare practitioners understand the patient's health history and make educated judgments.

3. Clinical Notes: EHRs contain clinical notes collected by healthcare providers during patient contacts. These notes capture symptoms, observations, exams, diagnoses, and treatment

strategies. They give a detailed review of the patient's medical appointments.

4. Medication and Allergies: EHRs hold information regarding the drugs provided to the patient, including the dose, frequency, and duration. They also note any known allergies or bad responses to certain medicines.

5. Laboratory and Imaging data: EHRs integrate and show laboratory test data, such as blood tests, pathology reports, and radiology imaging findings. These data may be evaluated throughout time, facilitating in the monitoring of the patient's health state.

6. Immunization Records: EHRs include records of the patient's immunizations, including the kind of vaccine, date given, and any adverse responses. This information helps ensure patients get adequate vaccines and prevent repeat doses.

7. Clinical Decision Support: EHR systems commonly contain clinical decision support tools, which employ algorithms and rules to deliver warnings, reminders, and suggestions to healthcare practitioners. These technologies assist improve patient safety, increase clinical decision-making, and encourage evidence-based practice.

8. Interoperability: EHR systems strive to support interoperability, facilitating the safe exchange of patient data across multiple healthcare venues and systems. This enables for smooth information exchange across healthcare professionals, enhancing care coordination and continuity.

By using EHR systems, healthcare practitioners may access complete patient information, make educated clinical choices, improve care coordination, and better patient outcomes. EHRs serve a vital role in facilitating the digitalization and transformation of healthcare delivery.

5.1.2 Benefits and Challenges of EHR Implementation

Implementing Electronic Health Records (EHR) provides various benefits and advantages for healthcare organizations, clinicians, and consumers. However, there are other obstacles related with EHR deployment that need to be addressed. In this part, we will analyze the pros and problems of EHR deployment.

Benefits of EHR Implementation:

1. Enhanced Accessibility and Availability of Patient Information: EHRs give healthcare practitioners with quick access to a patient's medical information, independent of their location. This enhances the efficiency of healthcare delivery, particularly in emergency circumstances or while treating patients in diverse healthcare facilities.

2. Improved Care Coordination and Continuity: EHRs allow seamless information interchange between various healthcare providers, enabling greater care coordination and continuity. This decreases the possibility of medical mistakes, duplicate testing, and delays in treatment.

3. Enhanced Patient Safety: EHRs facilitate the integration of clinical decision support systems, which may deliver real-time warnings, reminders, and evidence-based recommendations to healthcare practitioners. This aids in detecting possible prescription mistakes, drug interactions, and allergy concerns, hence enhancing patient safety.

4. Increased Efficiency and Productivity: EHRs simplify administrative operations, automate documentation procedures, and reduce the need for paper-based record-keeping. This increases

workflow efficiency, eliminates paperwork, and enables healthcare practitioners to spend more time on patient care.

5. Data-driven Insights and Population Health Management: EHRs support the gathering and analysis of patient data, enabling healthcare organizations to develop insights into population health trends, illness patterns, and treatment results. This enables preventive interventions, public health efforts, and evidence-based decision-making.

Challenges of EHR Implementation:

1. Cost of Implementation and Maintenance: Implementing an EHR system might include considerable upfront expenditures, including infrastructure setup, software license, personnel training, and continuing maintenance. Smaller healthcare providers may experience financial hurdles in implementing EHRs.

2. Workflow Disruptions and Learning Curve: Transitioning from paper-based records to EHRs demands changes in workflows and may initially interrupt existing procedures. Healthcare providers and employees require time to adjust to the new system and learn how to successfully utilize EHR functions.

3. Data Security and Privacy Concerns: EHRs include sensitive patient information, therefore protecting data security and privacy is of highest significance. Healthcare businesses must employ comprehensive security measures, including encryption, access restrictions, and frequent audits, to safeguard patient data from unwanted access or breaches.

4. Interoperability and Data Exchange: Achieving smooth interoperability across various EHR systems and healthcare providers remains a problem. Variations in data formats, standards, and technological capabilities might limit efficient data interchange and care coordination.

AI Revolution in Medicine, The Future of Healthcare

5. User happiness and Burnout: The usability of EHR systems plays a vital influence in user happiness among healthcare practitioners. Poorly designed or convoluted EHR interfaces may cause to provider irritation, cognitive fatigue, and possible burnout.

Addressing these difficulties needs careful planning, stakeholder participation, training, and continuous assistance from healthcare organizations and EHR suppliers. By proactively addressing these difficulties, healthcare organizations may maximize the advantages of EHR deployment and enhance patient care.

5.2 AI Applications in EHR

5.2.1 Data Mining and Pattern Recognition

AI has changed the study of electronic health records (EHR) by employing powerful data mining and pattern recognition algorithms. In this part, we will study the uses of AI in data mining and pattern detection inside EHR systems.

Data mining includes the identification and extraction of significant patterns, insights, and information from huge amounts of EHR data. AI algorithms can process and analyze organized and unstructured data inside EHRs, allowing healthcare clinicians to extract valuable information for different reasons.

1. Predictive Analytics: AI systems can evaluate previous EHR data to uncover patterns and trends that might predict future health outcomes. For example, machine learning models may forecast the chance of a patient having a given illness based on their demographic information, medical history, and lifestyle variables. This may aid healthcare practitioners in early intervention, preventative care, and tailored treatment planning.

2. Clinical Decision Support: AI algorithms may evaluate EHR data to provide clinical decision support, delivering evidence-based suggestions to healthcare clinicians. By combining medical

knowledge and patient-specific data, AI systems may recommend suitable diagnostic tests, treatment choices, and drug doses. This may increase the precision of clinical decision-making, minimize mistakes, and boost patient safety.

3. Disease Surveillance and Outbreak Detection: AI algorithms can evaluate EHR data from huge populations to detect disease outbreaks and track the spread of infectious illnesses. By finding trends and abnormalities in the data, AI systems may give early warnings, promote public health programs, and allow prompt responses to impending outbreaks.

4. Adverse Event Detection: AI algorithms may examine EHR data to detect probable adverse events connected to pharmaceutical usage, medical procedures, or other healthcare interventions. By analyzing trends and correlations within the data, AI systems may warn healthcare personnel to possible hazards, allowing for prompt intervention and avoidance of bad outcomes.

5. Patient Risk Stratification: AI algorithms can scan EHR data to stratify patients into risk groups based on their health conditions, comorbidities, and other pertinent criteria. This may assist healthcare practitioners prioritize therapies, distribute resources effectively, and enhance population health management.

6. Clinical Research and Evidence Generation: AI algorithms may scan aggregated EHR data from different sources to provide real-world evidence for clinical research. By finding trends in treatment results, illness development, and responsiveness to therapies, AI systems may help to the creation of evidence that supports clinical trials, comparative effectiveness studies, and healthcare policy choices.

To successfully employ AI in data mining and pattern recognition inside EHR systems, healthcare companies must assure data quality, privacy protection, and adherence to ethical principles.

AI Revolution in Medicine, The Future of Healthcare

Collaboration between healthcare practitioners, data scientists, and AI specialists is vital to construct powerful AI models that can find significant insights from EHR data while respecting patient privacy and confidence.

5.2.2 Predictive Analytics and Early Disease Detection

One of the primary uses of AI in electronic health records (EHR) is predictive analytics for early illness identification. By examining enormous amounts of patient data, AI systems may uncover patterns, risk factors, and indicators that signal the possibility of getting specific illnesses. This allows healthcare practitioners to intervene early, leading to more effective therapies and better patient outcomes.

1. Risk Stratification: AI algorithms may scan EHR data to stratify patients into distinct risk groups based on their demographic information, medical history, genetic variables, lifestyle decisions, and other pertinent data. By identifying patients at high risk for certain illnesses, healthcare practitioners may proactively engage with preventative measures, screening programs, and tailored therapies.

2. Disease Prediction: AI models can learn from previous EHR data to forecast the chance of acquiring particular illnesses. These models evaluate numerous characteristics such as age, gender, family history, biomarkers, and clinical data to provide individualized risk ratings. For example, AI systems can forecast the chance of getting cardiovascular illnesses, diabetes, cancer, or neurodegenerative disorders.

3. Early Warning Systems: By continually monitoring and analyzing EHR data, AI systems may discover early warning indicators of illness development or aggravation. This may be especially effective in chronic illnesses when early management might reduce complications and enhance results. For instance, AI

may identify changes in blood glucose levels in diabetic patients or decline in lung function in persons with respiratory disorders.

4. Alert Systems: AI may be applied to construct alert systems that inform healthcare professionals when particular clinical problems or risk factors are found in EHR data. For example, if an EHR record shows a possible drug-drug interaction or a patient's laboratory values depart from the predicted range, AI algorithms may trigger warnings to enable prompt intervention and avert bad occurrences.

5. Population Health Management: AI-powered predictive analytics may assist healthcare companies uncover health trends and patterns among populations. By evaluating aggregated EHR data, AI algorithms may identify illness clusters, estimate population health risks, and enable targeted public health actions. This permits proactive control of population health and resource allocation.

It is vital to emphasize that the use of predictive analytics in EHR needs careful evaluation of data quality, privacy, and ethical problems. Adequate data governance, patient permission, and compliance with legislation are necessary to guarantee the appropriate and ethical use of predictive analytics. Additionally, healthcare professionals should regularly verify and develop AI models to boost their accuracy and dependability.

By exploiting the potential of predictive analytics and early illness identification, AI in EHR may transform healthcare by moving the emphasis from reactive to proactive and preventative treatment, eventually improving patient outcomes and decreasing healthcare costs.

5.2.3 Clinical Decision Support Systems (CDSS)

Clinical Decision Support Systems (CDSS) powered by AI play a significant role in using electronic health records (EHR) to enhance clinical decision-making and improve patient care. CDSS combine patient-specific data from EHR with medical knowledge and algorithms to give physicians with evidence-based recommendations, warnings, and advice at the point of treatment.

1. Evidence-Based Guidelines: CDSS may incorporate known clinical guidelines and procedures into the EHR process. By evaluating patient data in real-time, CDSS may give physicians with appropriate guidelines and suggestions to assist diagnosis, treatment, and follow-up care. This guarantees that professionals have access to the latest evidence-based information, leading to more standardized and effective treatment.

2. Drug Interaction and Allergy Checking: AI-powered CDSS can discover possible drug interactions and allergies by evaluating patient prescription lists and medical history. When prescribing pharmaceuticals, the system may create alerts and cautions if there are possible interactions or allergies, helping physicians make better prescription selections and prevent adverse occurrences.

3. diagnosis Support: CDSS may help in the diagnosis process by assessing patient symptoms, medical history, and test findings. By comparing the patient's data to a massive library of medical knowledge and trends, the system may create differential diagnoses, recommend further testing or imaging, and give physicians with decision assistance to obtain accurate and rapid diagnoses.

4. Risk Stratification and Prognostic Tools: AI algorithms inside CDSS may examine patient data to estimate the risk of developing specific illnesses or consequences. This helps doctors identify patients who may benefit from specific therapies or preventative measures. Additionally, CDSS may give prognostic tools that assess the expected outcomes of certain treatments or interventions, supporting doctors in tailored treatment planning.

5. Clinical Workflow Optimization: CDSS may optimize clinical processes by automating repetitive operations, such as documentation, order input, and result interpretation. This eliminates administrative complexity and enables physicians to concentrate more on patient care. CDSS may also assist select and triage patient cases based on urgency and severity, optimizing resource allocation and enhancing efficiency.

It is necessary to address possible problems while deploying CDSS, such as system integration, data interoperability, and user acceptability. Effective training and education for doctors are needed to enable optimal usage of CDSS and to address problems linked to dependence on technology.

By integrating AI inside CDSS, healthcare practitioners may benefit from better clinical decision-making, fewer medical mistakes, higher adherence to guidelines, and enhanced patient safety. The combination of CDSS with EHR produces a strong synergy that promotes evidence-based practice and allows tailored, patient-centered care.

5.3 Natural Language Processing (NLP) in EHR
5.3.1 Text Mining and Information Extraction

Natural Language Processing (NLP) is an area of AI that focuses on the interface between computers and human language. In the context of electronic health records (EHR), NLP plays a crucial role in extracting and interpreting essential information from unstructured clinical language, such as physician's notes, radiology reports, and discharge summaries.

1. content Mining: NLP approaches allow the mining of vast amounts of unstructured clinical content inside EHR. By evaluating free-text documents, NLP algorithms may discover and extract crucial information, such as patient demographics, medical

problems, prescriptions, treatments, and test findings. This information may then be formatted and incorporated into the EHR system, making it conveniently available for clinical decision-making and research reasons.

2. Information Extraction: NLP algorithms can extract particular information from clinical tales and provide suitable semantic labels. For example, NLP can detect and extract pharmaceutical names, doses, frequencies, and routes of administration from physician's notes. This allows the automatic populating of prescription lists, improves medication reconciliation procedures, and helps drug safety programs.

3. Clinical Coding and categorization: NLP may aid in automated coding and categorization of clinical information inside EHR. This covers activities such as assigning diagnostic codes (e.g., ICD-10) and procedure codes (e.g., CPT codes) based on the information provided in clinical text. NLP algorithms can reliably detect important clinical ideas and translate them to suitable codes, decreasing human coding labor and boosting coding accuracy.

4. Clinical Documentation Improvement: NLP may assist in increasing the quality and completeness of clinical documentation. By evaluating clinical narratives, NLP algorithms may detect missing or incomplete information and create prompts or recommendations to aid doctors in obtaining crucial material. This boosts the accuracy and comprehensiveness of the EHR, leading to greater communication and collaboration among healthcare professionals.

5. Clinical Research and Population Health: NLP facilitates the extraction of data for clinical research and population health management from EHR. By evaluating clinical narratives, NLP algorithms may identify patients with certain diseases, evaluate eligibility for research projects, and simplify cohort identification.

AI Revolution in Medicine, The Future of Healthcare

NLP also plays a vital role in extracting structured data for clinical registries and population health analytics, providing insights into illness trends, treatment results, and healthcare usage.

However, hurdles abound in NLP implementation, including the variety of clinical language, the necessity for domain-specific training data, and maintaining the privacy and security of patient information. Ongoing research and development activities in NLP continue to solve these problems and expand the capabilities of NLP in EHR.

By integrating NLP inside EHR, healthcare companies may extract useful insights from unstructured clinical text, improve clinical decision support, boost coding accuracy, and support clinical research and population health efforts. NLP offers enormous promise in maximizing the usage of EHR data and revolutionizing healthcare delivery and research.

5.3.2 Clinical Documentation Improvement

Clinical documentation plays a significant part in healthcare, as it acts as a record of patient interactions, diagnosis, treatments, and results. Accurate and complete clinical documentation is critical for successful communication, care coordination, invoicing, and research. However, healthcare professionals sometimes confront difficulty in recording comprehensive and accurate clinical information inside electronic health records (EHRs). This is where AI, notably Natural Language Processing (NLP), may aid to clinical documentation improvement initiatives.

1. Automated Documentation Review: NLP algorithms may evaluate clinical documentation inside EHRs to detect gaps, discrepancies, and opportunities for improvement. By evaluating clinical narratives, NLP algorithms may identify missing or incomplete information, such as allergies, medication reconciliation, or pertinent patient history. This automated review

method helps healthcare practitioners verify that all relevant information is captured, leading to more accurate and thorough patient records.

2. Clinical Documentation Guidance: NLP may give real-time prompts or ideas to aid healthcare practitioners throughout the documentation process. As physicians enter information into the EHR, NLP algorithms may scan the language and make suggestions for more details or particular documentation items that should be included. These prompts may assist healthcare practitioners record the relevant information, increase documentation completeness, and promote proper coding and invoicing.

3. Coding Assistance: NLP may aid in clinical coding by automatically extracting important clinical ideas from documentation and mapping them to suitable coding systems, such as ICD-10 or CPT codes. By evaluating clinical narratives, NLP algorithms may discover diagnoses, procedures, and other pertinent information, minimizing the strain on coders and assuring coding accuracy. This not only enhances billing and reimbursement procedures but also promotes data quality and analytics.

4. Quality Metrics and Compliance: NLP may evaluate clinical paperwork to determine compliance with quality metrics and documentation rules. By assessing paperwork against established norms and guidelines, NLP algorithms may detect areas of non-compliance and offer feedback to healthcare professionals. This assists in improving documentation processes, satisfying regulatory standards, and raising the overall quality of service.

5. Clinical Documentation Integrity Auditing: NLP may be applied to undertake retrospective audits of clinical documentation to detect possible errors, such as inconsistencies, inaccuracies, or missing information. By examining a huge number of patient information, NLP systems may highlight records that need

additional evaluation or correction. This helps healthcare organizations verify the integrity and correctness of their clinical record, minimizing the risk of medical mistakes and enhancing patient safety.

It is vital to remember that AI-driven clinical documentation enhancement should always be utilized as a supporting tool, with healthcare personnel ultimately accountable for the quality and completeness of the record. Collaboration between AI systems and healthcare practitioners is vital to utilize the advantages of NLP while keeping the human touch and clinical judgment in the documentation process.

By employing NLP for clinical documentation enhancement, healthcare organizations may increase the quality and integrity of their patient records, improve communication and care coordination, streamline billing procedures, and enable research and analytics. AI-assisted clinical documentation enhancement offers the potential to expedite processes, minimize documentation burdens, and ultimately improve patient care.

5.3.3 Automated Coding and Billing

Coding and billing in healthcare are complicated operations that need great attention to detail and adherence to specialized coding systems and norms. The emergence of AI, especially Natural Language Processing (NLP), has enabled breakthroughs in automating coding and billing processes inside electronic health records (EHRs). Here, we study the uses and advantages of AI in automated coding and billing.

1. Code Suggestion and Validation: NLP algorithms may evaluate clinical data inside EHRs to extract key clinical ideas and offer suitable codes for diagnoses, treatments, and services. By integrating machine learning methods and medical coding knowledge bases, these algorithms may give real-time

recommendations to healthcare practitioners while they record patient interactions. This helps guarantee precise and consistent coding, minimizing the likelihood of coding mistakes and enhancing reimbursement accuracy.

2. Streamlined Billing procedures: AI-powered solutions may automate the coding and billing procedures, decreasing human labor and enhancing efficiency. NLP algorithms can extract essential information from clinical paperwork and produce billing codes automatically. This automation improves the billing system, minimizes administrative effort, and speeds the revenue cycle.

3. Revenue Optimization: AI may aid in improving revenue by recognizing missed billing possibilities and probable undercoding or overcoding occurrences. NLP algorithms may examine clinical paperwork to guarantee that all billable services and procedures are collected and appropriately coded. By spotting any possible conflicts between the recorded clinical information and the billed codes, AI systems may assist healthcare companies enhance their revenue collection.

4. Compliance and Audit Support: Automated coding and billing systems may contain compliance checks to verify conformity to coding and billing standards and norms. NLP algorithms can check that the recorded clinical information supports the billable codes, lowering the possibility of fraudulent or incorrect billing. These tools may also assist retrospective audits by evaluating coding and billing trends to discover any inconsistencies or areas for improvement.

5. Documentation Improvement: AI-powered coding and billing systems may give feedback to healthcare practitioners about documentation completeness and correctness. By examining the clinical paperwork, NLP algorithms may discover locations where further details or particular information are required to enable

acceptable coding and billing. This feedback loop helps improve documentation processes, leading to more accurate coding, better reimbursement, and greater compliance.

While automated coding and billing systems provide substantial benefits, it's necessary to guarantee constant monitoring and quality control. Human monitoring and validation are required to maintain the correctness and integrity of the coding and invoicing process. Collaboration between AI systems and coding/billing specialists is vital to establish a balance between automation and human knowledge.

AI-driven automated coding and billing have the potential to boost accuracy, expedite operations, and improve revenue collection in healthcare businesses. By employing NLP algorithms to extract and analyze clinical information, these systems may eliminate coding mistakes, streamline billing procedures, and assist compliance with coding and billing standards.

5.4 AI-Enabled Clinical Workflow Optimization

5.4.1 Intelligent Scheduling and Resource Allocation

Efficient scheduling and resource allocation are critical for streamlining clinical processes and guaranteeing timely and successful healthcare delivery. AI-enabled systems provide potential solutions to solve the issues involved with managing complicated clinical schedules and allocating resources efficiently. In this part, we study the uses and advantages of AI in intelligent scheduling and resource allocation.

1. optimum Appointment Scheduling: AI algorithms may examine different parameters, such as patient preferences, healthcare provider availability, and resource limits, to develop optimum appointment schedules. By evaluating many criteria and applying machine learning approaches, these algorithms may cut waiting times, alleviate scheduling conflicts, and increase overall

patient satisfaction. AI can also assist handle appointment cancellations and rescheduling by automatically changing the calendar and filling empty slots effectively.

2. Resource use and Capacity Planning: AI can assess historical data, patient demand trends, and resource availability to maximize resource use and plan capacity efficiently. By anticipating patient loads, AI systems may assist healthcare businesses deploy employees, equipment, and facilities effectively. This guarantees that the necessary resources are accessible at the right moment, avoiding bottlenecks and increasing the usage of resources.

3. Real-Time Workflow Monitoring: AI can continually monitor the status of healthcare processes and detect bottlenecks or delays in real-time. By combining data from numerous sources, like as EHRs, monitoring devices, and administrative systems, AI algorithms may give insights into the present state of patient care operations. This helps healthcare practitioners to proactively monitor processes, make educated choices, and take appropriate measures to preserve the smooth flow of operations.

4. Dynamic Resource Allocation: AI systems can dynamically distribute resources depending on real-time needs and changing priorities. For example, in emergency departments or operating rooms, AI may improve resource allocation by assessing parameters such as patient acuity, urgency, and resource availability. By continually evaluating data and altering resource allocation, AI systems may assist healthcare practitioners give timely treatment and react efficiently to unanticipated occurrences or catastrophes.

5. Workflow Optimization and Predictive Analytics: AI can evaluate enormous amounts of data, including patient records, clinical guidelines, and treatment regimens, to detect trends and enhance clinical processes. By employing machine learning algorithms, AI systems may recommend process changes, detect

inefficiencies, and simplify processes for improved patient outcomes. Predictive analytics may also assist estimate future demand and resource demands, allowing proactive planning and resource allocation.

It is vital to stress that AI-enabled scheduling and resource allocation systems should be created in partnership with healthcare professionals and stakeholders. They should examine clinical knowledge, patient preferences, and ethical concerns to ensure that decision-making corresponds with high-quality treatment and patient-centered concepts.

AI-driven intelligent scheduling and resource allocation have the potential to boost efficiency, maximize resource use, and improve patient happiness in healthcare settings. By employing AI algorithms to evaluate data, estimate demand, and improve processes, these systems may assist healthcare companies offer timely and efficient treatment while guaranteeing optimum usage of resources.

5.4.2 Automated Documentation and Reporting

Accurate and accurate documentation is vital in healthcare for preserving patient records, disseminating information among healthcare practitioners, and guaranteeing correct compensation. However, the process of documenting may be time-consuming and prone to mistakes. AI technology provides solutions to automate and expedite documentation and reporting activities, enhancing productivity and decreasing the stress on healthcare workers. In this part, we study the uses and advantages of AI in automated documentation and reporting.

1. speech Recognition and Transcription: AI-powered speech recognition technology can turn spoken language into text, enabling healthcare practitioners to dictate their findings, observations, and treatment recommendations straight into electronic health records

(EHRs). This removes the need for human transcribing and considerably decreases the time and effort necessary for documentation. Voice recognition systems may also adapt to the specialized language and vocabulary used in healthcare, delivering accurate and contextually appropriate documentation.

2. Natural Language Processing (NLP): NLP approaches allow the extraction and analysis of relevant information from unstructured clinical material. AI algorithms can analyse clinical notes, medical reports, and other textual data to identify essential clinical ideas, extract significant information, and arrange it into standardized forms. This not only automates the documenting process but also permits data mining, decision support, and clinical research based on the retrieved information.

3. Clinical Coding and categorization: AI may aid with automated coding and categorization of medical diagnoses, procedures, and therapies. By examining clinical data, AI algorithms may properly assign suitable codes according to conventional coding systems, such as ICD-10 (International Classification of Diseases) or CPT (Current Procedural Terminology). Automated coding not only increases coding accuracy but also simplifies the billing and payment process, minimizing administrative hassles and possible mistakes.

4. Quality Assurance and Compliance: AI algorithms may examine clinical paperwork for quality assurance purposes, assuring compliance with documentation standards and regulatory criteria. They may identify missing or incomplete information, find discrepancies, and offer real-time feedback to healthcare practitioners. This helps increase documentation accuracy and completeness, which is critical for efficient communication, continuity of care, and satisfying legal and regulatory duties.

5. Automated Reporting and Summarization: AI can create automated reports and summaries based on organized clinical data. By evaluating patient data, test findings, and treatment information, AI algorithms may create brief and standardized reports, such as discharge summaries, progress notes, or referral letters. Automated reporting saves time, standardizes documentation formats, and promotes communication among healthcare practitioners, encouraging greater care coordination and patient safety.

It is vital to remember that although AI may automate different elements of documentation and reporting, healthcare personnel remain responsible for validating and analyzing the produced information to guarantee correctness and clinical validity. Collaboration between AI systems and human professionals is vital to achieve the correct balance between automation and clinical control.

AI-driven automated documentation and reporting have the potential to simplify clinical procedures, minimize administrative load, and increase the quality and efficiency of healthcare documentation. By utilizing speech recognition, NLP, and automated coding, these technologies may boost documentation accuracy, allow data-driven decision-making, and free up critical time for healthcare practitioners to concentrate on patient care.

5.4.3 Workflow Streamlining and Efficiency Improvement

Efficient and well-organized processes are critical in healthcare to maximize patient care delivery, eliminate mistakes, and increase overall operational efficiency. AI technology provides chances to expedite and improve processes, automating mundane operations, offering decision assistance, and boosting communication and coordination among healthcare personnel. In this part, we study the uses and advantages of AI in process streamlining and efficiency enhancement.

1. Task Automation: AI can automate repetitive and time-consuming processes, enabling healthcare personnel to concentrate on more complicated and vital duties. For example, AI-powered chatbots or virtual assistants may handle appointment scheduling, patient registration, and basic queries, freeing up administrative staff's time and lowering wait times for patients. Robotic process automation (RPA) may automate data input, document management, and other administrative operations, reducing human mistakes and speeding up procedures.

2. Intelligent Resource Allocation: AI algorithms can assess patient data, resource availability, and operational restrictions to improve resource allocation. For example, AI may aid in scheduling operations, treatments, and consultations, taking into account criteria like surgeon availability, operating room availability, and patient priority. By intelligently distributing resources, AI can minimize wait times, increase resource utilization, and boost patient flow across healthcare facilities.

3. Decision Support Systems: AI-powered decision support systems (DSS) may aid healthcare practitioners in making well-informed choices at different stages along the treatment route. DSS may assess patient data, medical literature, and clinical guidelines to give evidence-based recommendations for diagnosis, therapy, and follow-up care. By incorporating AI algorithms into clinical processes, healthcare professionals may benefit from timely and customized decision assistance, leading to better clinical outcomes and decreased variability in treatment.

4. Communication and cooperation: AI technology, such as secure messaging systems or digital platforms, may promote smooth communication and cooperation among healthcare practitioners. These systems may give real-time notifications, alerts, and updates, allowing efficient coordination of care and effective sharing of essential information. AI-powered natural language processing

(NLP) may aid in extracting and summarizing significant information from medical records, easing information exchange and boosting multidisciplinary cooperation.

5. Predictive Analytics: AI systems can evaluate vast amounts of healthcare data, including patient records, physiological signals, and medical imaging, to uncover patterns and trends that might enable proactive decision-making. Predictive analytics may assist healthcare companies predict patient requirements, identify at-risk groups, and allocate resources appropriately. For example, AI can forecast patient readmissions, identify patients at high risk for developing problems, or improve bed management based on projected patient flow.

6. Continuous Learning and Improvement: AI systems may learn from real-world data and user interactions, continually improving their performance over time. By monitoring results and feedback, AI algorithms may adjust and modify their suggestions, ensuring that the workflow optimization tactics are matched with developing healthcare demands and goals.

While AI has enormous potential to simplify processes and enhance efficiency, it is crucial to address the ethical implications and maintain a balance between automation and human engagement. Close coordination between healthcare practitioners, AI specialists, and policymakers is vital to enable the appropriate adoption of AI technology in healthcare operations.

AI-driven workflow simplification and efficiency improvement may lead to greater patient happiness, decreased costs, and improved healthcare outcomes. By automating chores, giving decision assistance, and enabling efficient communication, AI technology may enhance healthcare delivery, allowing healthcare personnel to concentrate on delivering high-quality, patient-centered care.

AI Revolution in Medicine, The Future of Healthcare

5.5 AI and Interoperability of EHR Systems
5.5.1 Data Standardization and Integration

Interoperability, the capacity of various healthcare systems and apps to share and utilize data smoothly, is a vital component of electronic health records (EHR) systems. AI may play a crucial role in strengthening the interoperability of EHR systems by tackling difficulties related to data standards and integration. In this part, we discuss the uses and advantages of AI in increasing interoperability in EHR systems.

1. Data Standardization: AI can assist solve the problem of data heterogeneity by standardizing and harmonizing health data from multiple sources. Natural language processing (NLP) methods can extract structured data from unstructured clinical notes, facilitating interoperability across systems that employ various data formats. AI may also aid in mapping and converting data from different sources to a single data model or standard, enabling data interchange and integration.

2. Data Integration: AI approaches, including as machine learning and data fusion, may combine data from numerous EHR systems, clinical databases, and other sources to offer a holistic perspective of patient health. By collecting and analyzing varied information, AI systems may develop insights, uncover trends, and enhance clinical decision-making. For example, AI may combine patient demographic data, medical history, laboratory results, and imaging findings to offer a holistic assessment of the patient's health state.

3. Semantic Interoperability: AI can enhance the semantic interoperability of EHR systems by boosting the comprehension and interpretation of clinical data. Natural language processing and semantic modeling approaches may extract meaning from clinical narratives, ontologies, and medical terminologies. AI algorithms can

assess the context and connections within healthcare data, allowing more accurate and precise data integration and interpretation.

4. Data sharing and Communication: AI technology can promote safe and efficient data sharing across various EHR systems and healthcare organizations. AI-powered data integration systems and interoperability frameworks may allow smooth communication and data exchange, ensuring that essential patient information is available when and when it is required. AI can also enhance data governance and privacy protection, ensuring that data sharing conforms with legal regulations and patient permission.

5. Clinical Decision Support: AI-driven clinical decision support systems (CDSS) may exploit interoperable EHR data to give timely and individualized suggestions to healthcare providers. By integrating with EHR systems, AI algorithms can access patient-specific data, including medical history, laboratory results, and medication records, to generate evidence-based treatment guidelines, alert providers to potential drug interactions or adverse events, and support care coordination across different healthcare settings.

6. Population Health Management: AI can exploit interoperable EHR data to assess population-level health patterns, identify high-risk populations, and improve healthcare budget allocation. By combining and analyzing data from numerous EHR systems, AI algorithms may produce insights into illness prevalence, treatment results, and healthcare usage trends. This information may guide public health interventions, promote population health management programs, and enable evidence-based policy-making.

Promoting interoperability in EHR systems using AI-driven technologies may boost data interchange, promote collaborative treatment, and improve healthcare outcomes. However, concerns like as data privacy, security, and governance need to be addressed to enable appropriate data sharing and maintain patient confidentiality. Collaboration among healthcare institutions, technology suppliers, and regulatory authorities is vital to define interoperability standards, build interoperable AI solutions, and promote a culture of data sharing and collaboration in healthcare.

5.5.2 Semantic Interoperability and Data Exchange

Semantic interoperability is a fundamental feature of EHR systems that includes the capacity to share and analyze data in a meaningful manner across various healthcare systems and organizations. AI may play a crucial role in promoting semantic interoperability and allowing efficient data interchange. In this part, we discuss the uses and advantages of AI in attaining semantic interoperability and facilitating smooth data sharing in EHR systems.

1. Semantic Data Mapping: AI tools, including as natural language processing (NLP) and machine learning, may assist the mapping of data items and ideas from one EHR system to another. By examining the structure and content of clinical data, AI algorithms may detect and align related data items, features, and codes, assuring semantic consistency and compatibility. This mapping procedure permits the transmission of data with proper meaning and interpretation across multiple systems.

2. Clinical Terminology and Ontologies: AI can promote semantic interoperability by exploiting clinical terminologies and ontologies. AI systems can extract and correlate pertinent medical ideas, codes, and terminologies from clinical data. By adopting standardized terminologies such as SNOMED CT or LOINC, AI may facilitate the uniform depiction and sharing of clinical information. Additionally, AI may aid in establishing and maintaining ontologies that capture the connections and meanings

of medical concepts, thus boosting the interoperability of EHR systems.

3. Data Validation and Quality enhancement: AI approaches may aid in data validation and quality enhancement during data sharing. AI algorithms may assess incoming data for completeness, correctness, and consistency, discovering flaws or anomalies that may impede interoperability. By automatically evaluating and cleaning the sent data, AI helps guarantee that the data being shared is trustworthy and complies to set standards.

4. Data Harmonization and Integration: AI can simplify the harmonization and integration of data from disparate EHR systems by detecting common data items and resolving semantic disparities. AI algorithms may evaluate and convert data from multiple sources into a single format or data model, promoting smooth integration and understanding. This harmonization allows healthcare practitioners to access and aggregate patient data from multiple sources, facilitating full and holistic patient care.

5. Clinical Decision Support and Care Coordination: AI-driven clinical decision support systems (CDSS) may exploit semantically compatible EHR data to deliver real-time advice and suggestions to healthcare providers. By connecting with EHR systems, AI algorithms may access patient data across multiple healthcare settings, simplifying care coordination and promoting evidence-based decision-making. Semantic interoperability guarantees that important patient information, such as medical history, allergies, and prescriptions, is correctly understood and used by CDSS.

6. Interoperability Standards and Frameworks: AI may contribute to the creation and implementation of interoperability standards and frameworks that improve semantic interoperability in EHR systems. AI technology may aid in the construction and maintenance of interoperability requirements, data models, and terminology

mappings. By automating the process of standard creation and updating, AI helps speed the adoption of semantic interoperability standards and decreases the load on healthcare organizations.

Achieving semantic interoperability and allowing efficient data interchange in EHR systems needs coordination among healthcare stakeholders, standards groups, and technology suppliers. The use of AI in semantic data mapping, clinical terminologies, data validation, and harmonization helps to seamless information interchange, better care coordination, and increased patient outcomes. However, it is necessary to address privacy, security, and permission issues to enable responsible data exchange and maintain patient confidentiality in the context of semantic interoperability.

5.5.3 Improving Continuity of Care and Patient Engagement

AI technologies have the potential to dramatically increase continuity of treatment and patient involvement within the setting of electronic health records (EHR). In this section, we investigate how AI may be exploited to enhance the smooth flow of treatment and enable patients to actively engage in their healthcare experience.

1. Care Coordination and Transition Management: AI can help care coordination by evaluating patient data across multiple EHR systems, detecting care gaps, and supporting seamless transitions between healthcare providers. By applying machine learning algorithms, AI can detect possible difficulties or adverse events, notifying healthcare professionals and allowing preventive treatments. This guarantees that vital information is transmitted properly and efficiently, avoiding mistakes and increasing the continuity of treatment.

2. Personalized Care Plans: AI may aid in the formulation of personalized care plans by evaluating patient data from EHR systems and producing customised suggestions. Machine learning algorithms may assess historical patient data, treatment results, and

evidence-based recommendations to produce personalised care plans that account particular patient requirements, preferences, and clinical considerations. This fosters patient-centered care and facilitates collaborative decision-making between healthcare practitioners and patients.

3. Patient Education and Empowerment: AI-powered EHR systems may empower patients by providing them with individualized health information, educational materials, and self-management tools. Natural language processing (NLP) approaches may allow EHR systems to extract significant information from clinical notes and provide it to patients in an intelligible fashion. AI chatbots or virtual assistants may give patients with real-time instruction, answering inquiries, and delivering assistance. This improves patient involvement, enhances health literacy, and encourages active participation in controlling one's health.

4. Remote Monitoring and Telehealth: AI technology can support remote monitoring and telehealth services, providing continuous monitoring of patient health outside conventional healthcare facilities. AI algorithms can examine data from remote monitoring devices, wearables, and patient-reported outcomes to identify early symptoms of deterioration or changes in health condition. This data may be connected with EHR systems, giving healthcare clinicians with real-time insights and allowing prompt treatments. distant monitoring and telehealth technologies increase access to treatment, particularly for patients in distant places or those with restricted mobility.

5. Patient-Generated Health Data (PGHD): AI can help the integration and analysis of patient-generated health data into EHR systems. This includes data from wearable devices, mobile apps, and patient-reported outcomes. AI systems can analyse and interpret PGHD to give valuable insights to healthcare practitioners, allowing more thorough evaluations of patient health. Integrating PGHD into

EHR systems promotes the accuracy of patient information, facilitates tailored treatment, and allows proactive health management.

6. Health Information sharing and Portability: AI technology can allow the safe sharing and portability of health information across multiple EHR systems. AI algorithms can assess and standardize data formats, resolve semantic inconsistencies, and guarantee the safe transmission of medical information. This lets people to access their medical information across numerous healthcare providers, increasing continuity of treatment, and allowing patients to actively engage in decision-making.

While AI provides great prospects to enhance continuity of care and patient participation, it is vital to address privacy, security, and ethical problems. Safeguarding patient confidentiality, obtaining informed consent, and preserving data security are crucial in integrating AI inside EHR systems to better patient-centered care and empower people in controlling their health.

5.6 Enhancing EHR Data Quality with AI

5.6.1 Automated Data Cleansing and Validation

Accurate and high-quality data inside electronic health records (EHR) is vital for successful healthcare delivery, research, and decision-making. AI technology may play a crucial role in boosting EHR data quality by automating data cleaning and validation operations. In this part, we study how AI might increase the quality, completeness, and dependability of EHR data.

1. Data Cleansing: AI systems can automatically find and repair mistakes, inconsistencies, and anomalies in EHR data. Machine learning models may learn from trends and historical data to identify and fix typical data input problems, such as misspellings, typographical errors, or improper coding. By applying natural

language processing (NLP) methods, AI can also find and repair anomalies in free-text clinical notes, maintaining the integrity of the recorded material.

2. Data Validation: AI may aid in assessing the correctness and completeness of EHR data by comparing it against external sources and reference databases. For instance, AI systems may cross-reference patient demographics with government-issued identity information to verify precise identification and avoid duplication. Additionally, AI may employ clinical guidelines, medical literature, and best practices to assess the appropriateness of treatment regimens, prescription doses, and diagnostic codes reported in EHRs.

3. Data Standardization: EHR data frequently originate from multiple sources, resulting to variances in terminologies, coding systems, and data formats. AI may assist in standardizing EHR data by mapping various terminologies to common ontologies and coding systems. Machine learning algorithms can understand the links between distinct terminology and ideas, allowing precise mapping and standardization of clinical data. This fosters consistency and interoperability across multiple EHR systems, simplifying data sharing and analysis.

4. Real-Time Data Quality Monitoring: AI can continually evaluate data quality in real-time, alerting healthcare providers and administrators to any concerns. By evaluating data patterns and trends, AI algorithms may uncover abnormalities, missing data, or data input mistakes that may compromise the reliability and validity of EHR information. Real-time data quality monitoring enables for rapid interventions, corrections, and proactive efforts to protect data integrity.

5. Data Completeness and Contextualization: AI can examine structured and unstructured data inside EHRs to assure data

completeness and boost contextual understanding. NLP algorithms may extract valuable information from clinical notes, lab reports, and other unstructured sources to enhance structured data fields. AI algorithms may also combine data from other sources, like as wearable devices or patient-reported outcomes, to give a more holistic perspective of patient health. By boosting data completeness and context, AI offers a more comprehensive picture of patient situations and aids informed decision-making.

6. Continuous Learning and Improvement: AI systems may continually learn from fresh data and feedback to enhance data quality over time. As additional data becomes available, AI models may update and modify their algorithms to react to new trends, rules, and best practices. This repeated learning process boosts the accuracy and efficacy of data cleaning and validation, resulting to higher-quality EHR data.

It is vital to emphasize that AI algorithms are not a replacement for human skill and judgment. They should be utilized as tools to help healthcare personnel in assuring data quality rather than replacing their duties. Additionally, questions about patient privacy, permission, and transparency should be at the forefront when using AI technologies for boosting EHR data quality.

5.6.2 Quality Control and Error Detection

Maintaining high-quality electronic health record (EHR) data is critical for accurate and trustworthy healthcare information. AI may contribute to quality control and error detection procedures, helping to find and resolve inconsistencies, faults, and possible concerns inside EHR systems. In this part, we study how AI may enhance quality control and mistake detection in EHR data.

1. Data Consistency and Integrity: AI algorithms may examine EHR data to find discrepancies and inaccuracies in reported information. By comparing data across multiple fields and patient

records, AI can find anomalies and indicate possible concerns, such as contradicting diagnoses, inconsistent prescription records, or missing data. This helps healthcare professionals to repair problems, maintaining the integrity and consistency of the EHR data.

2. Anomaly Detection: AI can discover abnormal patterns or outliers within EHR data that may suggest mistakes or odd conditions. Machine learning algorithms may learn from previous data and find trends that depart from the norm. For example, AI may detect unexpectedly high or low values in laboratory test results or identify aberrant pharmaceutical prescriptions. Anomaly detection helps healthcare professionals uncover any mistakes or inconsistencies in EHR data and take necessary remedial steps.

3. mistake repair and Imputation: AI may aid in mistake repair and data imputation operations. When faults are found, AI systems may automatically recommend fixes based on set rules or learned patterns. For example, if a prescription dose is recorded wrongly, AI may advise the most probable adjustment based on prior data or known prescribing trends. Additionally, AI may impute missing data by assessing available information and generating informed guesses, assuring the completeness of the EHR records.

4. Data Validation versus Clinical standards: AI algorithms may check EHR data against known clinical standards and best practices to find any mistakes or deviations. By combining machine learning and NLP approaches, AI can extract essential information from clinical recommendations and correlate it with associated EHR data. For instance, AI may assess if prescription drugs comply with established treatment recommendations or whether the described symptoms correspond with recognized diagnosis criteria. This validation method helps to guarantee adherence to clinical standards and lowers the probability of mistakes in treatment and documentation.

5. Real-Time Error Monitoring: AI can continually scan EHR data in real-time, detecting any mistakes or discrepancies as they arise. By employing rule-based systems or machine learning models, AI algorithms may examine incoming data and compare it against established criteria. Real-time error monitoring provides fast detection and repair of mistakes, decreasing the possible effect on patient care and enhancing data quality.

6. Continuous Improvement: AI systems may learn from feedback and user interactions to continually enhance mistake detection and quality control methods. As healthcare professionals rectify mistakes and offer feedback, AI models may update and develop their algorithms to boost accuracy and efficacy. This recurrent learning approach helps to construct more robust mistake detection systems and improves overall data quality in EHR systems.

It is vital to stress that AI algorithms should be visible and explainable, enabling healthcare practitioners to comprehend and evaluate the ideas or conclusions made by the system. Additionally, human supervision and judgment remain vital in the quality control and mistake detection process, ensuring that AI outputs are thoroughly evaluated and validated by healthcare experts.

5.6.3 Data Governance and Privacy Protection

The integration of AI in electronic health records (EHR) brings forward crucial challenges surrounding data governance and privacy protection. As EHR systems handle sensitive patient information, it is vital to maintain effective data governance policies and controls to preserve patient privacy. In this part, we study how AI might help to data governance and privacy protection in the context of EHR.

1. Access Control and Authentication: AI may assist increase access control measures inside EHR systems. Facial recognition, fingerprint scans, or voice recognition technology may be connected

with authentication systems to guarantee that only authorized healthcare workers can access patient data. AI systems may also detect and flag abnormal access patterns, helping to identify possible unwanted access attempts and secure patient privacy.

2. Privacy-Preserving Data exchange: AI approaches such as federated learning and homomorphic encryption allow privacy-preserving data exchange. Federated learning enables AI models to be trained directly on decentralized EHR data without the requirement for data sharing, maintaining patient privacy. Homomorphic encryption enables calculations to be conducted on encrypted data, guaranteeing that sensitive information stays encrypted throughout AI analysis. These strategies facilitate cooperation and information exchange among healthcare practitioners while respecting patient privacy.

3. Anonymization and De-identification: AI can aid in the anonymization and de-identification of EHR data, minimizing the danger of patient re-identification. AI algorithms may automatically delete or obscure personally identifying information (PII) from EHR records while keeping the therapeutic value of the data. Techniques such as differential privacy may be applied to introduce noise or disturbances to the data, further safeguarding patient privacy.

4. Data Monitoring and Auditing: AI algorithms may monitor EHR data for possible privacy breaches or illegal access attempts. By examining access records and user activity patterns, AI can discover odd or suspicious actions that may suggest privacy violations. Additionally, AI may aid in reviewing data consumption and tracking the flow of information inside the EHR system, assuring compliance with privacy legislation and standards.

5. Compliance with Data Protection rules: AI can assist healthcare firms comply with data protection rules such as the General Data Protection Regulation (GDPR) and Health Insurance

Portability and Accountability Act (HIPAA). AI algorithms can automate the identification of sensitive information, enforce data handling regulations, and guarantee that data access and use correspond to legal requirements. By integrating AI-powered compliance monitoring and reporting solutions, healthcare firms may show their commitment to preserving patient privacy.

6. Ethical principles and Bias Mitigation: AI algorithms employed in EHR should be developed to minimize bias and preserve ethical principles. Care must be made to ensure that AI models do not perpetuate or magnify current biases in healthcare. Regular audits and fairness evaluations of AI algorithms may assist discover and eliminate any biases inherent in the system. Transparency and explainability of AI algorithms are vital to enable healthcare professionals and patients to understand how choices are made and identify any possible biases.

It is vital to develop clear regulations and procedures for data governance and privacy protection in AI-driven EHR systems. Collaborative efforts between healthcare institutions, politicians, and technology developers are vital to guarantee the appropriate and ethical use of AI while safeguarding patient privacy and data security.

5.7 Ethical Considerations in AI-Driven EHR

5.7.1 Privacy and Security of Patient Data

Privacy and security of patient data are crucial ethical issues when deploying AI-driven electronic health record (EHR) systems. As AI algorithms collect and analyze massive amounts of sensitive medical information, it is crucial to prioritize privacy and security to safeguard patient confidentiality and maintain confidence in healthcare systems. In this part, we investigate the ethical concerns connected to the privacy and security of patient data in AI-driven EHR.

AI Revolution in Medicine, The Future of Healthcare

1. Data Encryption and Access Control: AI-driven EHR systems should implement powerful encryption mechanisms to secure patient data both at rest and during transmission. Encryption helps guarantee that even if unauthorized parties get access to the data, they cannot decode its contents. Access control methods should be created to limit data access to authorized healthcare workers, with varying degrees of access rights depending on their roles and responsibilities.

2. Data reduction and Purpose restriction: To protect privacy norms, AI-driven EHR systems should adhere to the concepts of data reduction and purpose restriction. This implies that just the essential patient data needed for specified healthcare objectives should be collected, processed, and maintained. Unnecessary or unnecessary data should not be gathered or maintained to reduce privacy hazards.

3. permission and Patient Empowerment: Informed permission is vital when utilizing AI algorithms on patient data. Healthcare practitioners should ensure that patients understand how their data will be used and receive their informed agreement for data sharing and analysis. Transparent communication and patient education about the advantages and possible hazards connected with AI-driven EHR systems allow patients to make educated choices about their data.

4. Data De-identification and Anonymization: AI algorithms employed in EHR systems should contain strategies for data de-identification and anonymization. Personally identifiable information (PII) should be erased or obscured to avoid re-identification of people. Aggregated and anonymized data may still give significant insights for research and population health management while safeguarding individual privacy.

5. Data Breach Prevention and Response: AI-driven EHR systems should have strong procedures in place to prevent data breaches and swiftly react to any security events. Regular security audits, vulnerability assessments, and personnel training programs may assist detect and minimize possible hazards. In the case of a data breach, healthcare institutions should have mechanisms in place to inform impacted persons and take necessary steps to limit damage.

6. Third-Party Vendor Evaluation: When incorporating AI technology into EHR systems, healthcare institutions should carefully analyze the privacy and security standards of third-party suppliers. It is crucial to verify that these providers comply to industry standards and best practices for data security. Contracts and agreements should clearly describe the rights and obligations of all parties regarding data privacy and security.

7. Regulatory Compliance: AI-driven EHR systems must comply with appropriate data protection and privacy requirements, such as the General Data Protection Regulation (GDPR) or Health Insurance Portability and Accountability Act (HIPAA). Healthcare firms should remain current on developing legislation and ensure that their systems and processes fit with the appropriate requirements.

By addressing the privacy and security of patient data, healthcare companies may develop confidence among patients, maintain ethical standards, and promote a responsible and safe environment for AI-driven EHR systems. Collaboration among stakeholders, including healthcare professionals, politicians, technology developers, and patients, is vital in tackling privacy and security problems and ensuring the appropriate use of AI in healthcare.

5.7.2 Bias and Fairness in EHR Data Analysis

Bias and fairness are major ethical issues when deploying AI-driven electronic health record (EHR) systems. AI algorithms that evaluate EHR data have the potential to perpetuate or magnify existing biases contained in the data, leading to unjust results and inequities in healthcare delivery. In this part, we investigate the ethical concerns linked to prejudice and fairness in EHR data processing.

1. Data Bias and Representation: EHR data may be skewed owing to several variables, including demographic, socioeconomic, and structural biases existing in healthcare systems. It is vital to be aware of such prejudices and take actions to overcome them. Care should be made to ensure that the data utilized for training AI algorithms is representative of varied patient populations, including different age groups, genders, races, nationalities, and socioeconomic backgrounds.

2. Algorithmic Bias: AI algorithms employed in EHR data analysis might accidentally add bias if they are trained on biased data or if the algorithm itself includes intrinsic biases. Bias may show in several ways, such as unequal accuracy or performance among different demographic groups. Healthcare institutions should frequently examine their AI algorithms for possible biases and strive towards establishing fair and impartial models.

3. justice in Decision-Making: AI-driven EHR systems should try to promote justice in decision-making processes. This implies that the system should not prejudice or favor particular persons or groups based on protected characteristics such as race, gender, or age. Fairness may be accomplished by careful algorithm design, assessing the influence of variables and characteristics utilized in decision-making, and regularly monitoring and evaluating the system's performance for fairness.

4. Transparency and Explainability: Transparency and explainability of AI algorithms employed in EHR data analysis are vital to reduce bias and promote justice. Healthcare companies

should attempt to make the decision-making processes of AI systems visible and give explanations for the outcomes or suggestions they create. This enables healthcare providers and patients to understand how the algorithm arrived at its results and helps detect and fix any biases or unfairness.

5. Continuous Evaluation and Monitoring: Bias and fairness in EHR data analysis should be regularly checked and analyzed. This entails continual examination of the algorithm's performance, reviewing its influence on various patient groups, and assessing the fairness of the results. Regular audits and reviews should be done to detect and resolve any biases or unfairness that may occur over time.

6. Collaboration and Diversity: Collaboration among different stakeholders, including healthcare professionals, data scientists, ethicists, and patients, is vital in resolving bias and fairness problems. Involving a wide variety of viewpoints may assist detect and minimize biases, question assumptions, and promote fair and equitable healthcare procedures. It is crucial to encourage a multidisciplinary approach to guarantee that ethical problems linked to prejudice and fairness are effectively addressed.

Addressing prejudice and ensuring fairness in EHR data analysis needs constant work and cooperation. By supporting openness, ongoing review, and a dedication to diversity and inclusion, healthcare companies may strive towards establishing AI-driven EHR systems that deliver fair and equitable healthcare for all persons, regardless of their qualities or origins.

5.7.3 Informed Consent and Patient Autonomy

In the context of AI-driven electronic health records (EHR), the ethical ideal of informed consent and patient autonomy plays a key role. In this part, we address the necessity of informed consent and patient autonomy in relation to AI-driven EHR systems.

1. Informed Consent: Informed consent is a key ethical concept that assures patients have the right to be fully informed about the purpose, risks, benefits, and possible results of any medical intervention, including the usage of AI-driven EHR systems. Patients should be given with clear and intelligible information on how their data will be collected, maintained, and utilized, especially in the context of AI algorithms evaluating their EHR data. Obtaining informed consent recognizes and respects patients' autonomy and enables them to make educated choices regarding their treatment.

2. Transparency in Data Use: Healthcare companies should endeavor to be upfront about how patient data will be utilized in AI-driven EHR systems. Patients should have a clear knowledge of how their data will contribute to improving healthcare outcomes, such as boosting diagnosis, treatment, or research. Transparency also requires alerting patients about the possible hazards and restrictions connected with the use of AI algorithms in their treatment.

3. ability to Opt-Out: Patients should have the ability to opt-out of participation in AI-driven EHR systems if they desire to do so. Respecting patient autonomy involves accepting their right to determine how their personal health information is utilized. Healthcare institutions should establish explicit ways for patients to exercise their right to opt-out, ensuring that their choice is recognized without affecting the quality of their treatment.

4. Education and Empowerment: To protect patient autonomy, it is vital to educate patients about the advantages and drawbacks of AI-driven EHR systems. Healthcare institutions should give accessible and clear information that allows patients to make educated choices about their involvement. This involves educating patients on the possible uses of AI, the privacy and security mechanisms in place to secure their data, and their rights around data use.

AI Revolution in Medicine, The Future of Healthcare

5. Shared Decision-Making: AI-driven EHR systems should facilitate shared decision-making between healthcare practitioners and patients. Patients should be actively included in the decision-making process surrounding the employment of AI technology in their treatment. This entails involving patients in dialogues about the possible advantages, hazards, and consequences of AI-driven EHR systems, letting them to voice their choices, values, and concerns.

6. Data Security and Privacy: Respecting patient autonomy also demands maintaining the security and privacy of their health data. Healthcare institutions must employ comprehensive data protection procedures to preserve patient information and prevent unwanted access or breaches. Patients should be assured that their data will be treated discreetly and in line with relevant privacy laws and regulations.

By preserving the concepts of informed consent and patient autonomy, healthcare institutions may develop a foundation of trust with patients about the usage of AI-driven EHR systems. This develops a patient-centric approach that respects individual choices, interests, and beliefs, eventually leading to greater patient participation and happiness in their healthcare experience.

5.8 Legal and Regulatory Challenges in AI-Driven EHR

5.8.1 Compliance with Data Protection Laws

The integration of AI into electronic health records (EHR) brings up legal and regulatory difficulties pertaining to data protection and privacy. Healthcare businesses implementing AI-driven EHR systems must guarantee compliance with appropriate data protection legislation, such as the General Data Protection Regulation (GDPR) in the European Union or the Health Insurance Portability and Accountability Act (HIPAA) in the United States. Key factors include:

a) Data Minimization: Healthcare institutions should gather and handle just the essential data for the intended purpose. They should guarantee that AI algorithms access and utilize just the minimal amount of patient data necessary to accomplish the intended results. Unnecessary or excessive data collecting should be avoided.

b) Anonymization and De-identification: EHR data utilized for AI purposes should be adequately anonymized or de-identified to preserve patient privacy. Personal identifiers and sensitive information should be erased or encrypted to avoid re-identification.

c) Data Storage and Encryption: Adequate security measures should be in place to safeguard EHR data from unwanted access, loss, or breaches. Robust encryption mechanisms and secure storage systems should be deployed to preserve patient information.

d) permission Management: Healthcare organizations must guarantee that patient permission for data usage in AI-driven EHR systems is acquired and handled effectively. Consent procedures should conform with relevant legal requirements and should be explicit, detailed, and transparent.

5.8.2 Regulatory Framework for AI in Healthcare

The application of AI in healthcare, especially AI-driven EHR systems, needs a robust regulatory framework to handle possible hazards, assure patient safety, and encourage ethical behaviors. Regulatory agencies and politicians are currently attempting to establish standards and legislation related to AI in healthcare. Some major factors include:

a) Validation and Certification: Regulatory frameworks should create methods for the validation and certification of AI algorithms utilized in EHR systems. This guarantees that the algorithms fulfill specified quality criteria, are safe, effective, and dependable.

b) Algorithmic openness and Explainability: Regulations should support openness and explainability of AI algorithms in EHR systems. Healthcare companies should be able to give clear explanations of how AI algorithms arrive at their judgments or recommendations, allowing healthcare practitioners and patients to trust and appreciate the logic behind the AI-driven outputs.

b) Ethical principles: Regulatory frameworks should integrate ethical principles for the use of AI in healthcare. These recommendations may encompass elements such as prejudice and fairness, informed consent, privacy protection, and the ethical use of AI technology.

d) supervision and Monitoring: Regulatory authorities should create procedures for continual supervision and monitoring of AI-driven EHR systems to guarantee compliance with regulatory standards. This may require frequent audits, evaluations, and reporting to ensure that AI algorithms and EHR systems conform to the specified rules.

e) Collaborative Approach: Regulatory frameworks should foster cooperation between regulatory agencies, healthcare organizations, AI developers, and other stakeholders. This cooperation may enable the establishment of comprehensive standards, exchange of best practices, and alignment of regulatory activities across multiple jurisdictions.

It is vital for healthcare businesses to be aware about shifting legal and regulatory environments linked to AI-driven EHR systems. By proactively addressing regulatory requirements and participating in ethical practices, healthcare businesses may utilize the advantages of AI while assuring the safety of patient data and sustaining confidence in the healthcare ecosystem.

5.8.3 Liability and Accountability in AI-Enabled EHR Systems

The incorporation of AI into electronic health record (EHR) systems raises fundamental problems surrounding responsibility and accountability. As AI algorithms play a key role in decision-making and medical care, understanding the distribution of culpability becomes critical. Here are significant issues in managing responsibility and accountability in AI-enabled EHR systems:

a) Algorithm Developer Liability: The developers or makers of AI algorithms employed in EHR systems may be held accountable for any damage caused by algorithmic mistakes or shortcomings. They are responsible for guaranteeing the safety, dependability, and correctness of their algorithms.

b) Healthcare Provider Liability: Healthcare providers who depend on AI-driven EHR systems must exercise due diligence in utilizing and interpreting the AI-generated recommendations. They should be aware of the limits and possible hazards connected with AI technology and exhibit professional judgment in their decision-making.

c) Shared responsibility: In circumstances where AI algorithms contribute to decision-making processes, responsibility may be shared between the algorithm creator, healthcare provider, and the healthcare organization. The attribution of accountability may rely on criteria such as the degree of human engagement, the transparency of the AI system, and the foreseeability of any mistakes.

d) Data Integrity and Governance: Healthcare companies have a duty to assure the integrity and quality of the data supplied into AI-enabled EHR systems. This involves confirming the authenticity and trustworthiness of the data sources and adopting rigorous data governance policies to prevent mistakes and biases that may harm AI-generated outputs.

e) Regulatory Oversight: Regulatory frameworks should address responsibility and accountability in AI-enabled EHR systems. These frameworks may create criteria for responsibility distribution, enforce reporting of bad occurrences or incidents connected to AI algorithms, and specify requirements for transparency, explainability, and traceability of AI-driven decision-making.

f) Documentation and Auditing: Adequate documentation of the AI algorithms used in EHR systems, including their creation, validation, and modifications, is vital for accountability. Regular auditing and monitoring of AI-driven decision-making processes may assist uncover possible mistakes or biases and allow accountability evaluations.

g) Professional Standards and Guidelines: Professional organizations and associations should create and promote standards and guidelines for the appropriate use of AI in healthcare. These guidelines may clarify the professional expectations and responsibilities of healthcare practitioners in implementing AI-enabled EHR systems.

h) educated permission and Patient Education: Healthcare organizations should ensure that patients are well-educated about the use of AI technology in their treatment and acquire their informed permission when appropriate. Patients should be taught about the possibilities and limits of AI-enabled EHR systems and given the chance to ask questions and make informed choices.

Addressing responsibility and accountability in AI-enabled EHR systems involves a multi-stakeholder strategy encompassing regulators, healthcare organizations, AI developers, healthcare providers, and legal experts. Collaborative efforts may help set clear norms, encourage ethical practices, and guarantee that patients are safeguarded while benefitting from the breakthroughs of AI in healthcare.

AI Revolution in Medicine, The Future of Healthcare

CHAPTER 6: AI AND PATIENT CARE

6.1 Transforming Patient Care with AI

6.1.1 The Impact of AI on Patient Care Delivery

In recent years, the integration of AI technology in patient care has profoundly revolutionized healthcare delivery, altering the way patients get and experience healthcare services. The influence of AI on patient care is multi-faceted, including different elements of healthcare delivery.

AI has allowed more accurate and quick diagnosis, increasing patient outcomes and minimizing medical mistakes. Through modern machine learning algorithms and data analysis, AI can evaluate huge volumes of patient data, including medical records, imaging findings, and genetic information, to aid healthcare personnel in making accurate and efficient diagnoses. This has led to earlier discovery of illnesses, more tailored treatment regimens, and better patient prognosis.

Furthermore, AI has increased patient monitoring and management, allowing remote patient monitoring and telemedicine services. With the use of wearable gadgets and linked health technologies, AI algorithms can continually monitor vital signs, identify irregularities, and offer real-time notifications to healthcare practitioners. This enables for proactive therapies, early illness identification, and better chronic disease management, eventually boosting patient outcomes and quality of life.

AI has also played a crucial role in drug management and adherence. Intelligent drug distribution systems may aid patients with complicated prescription regimens by offering individualized reminders, dosage instructions, and medication reconciliation. AI algorithms can improve pharmaceutical regimens based on unique patient variables, such as age, weight, and comorbidities, assuring safer and more effective therapy.

In addition, AI has changed patient involvement and education. Virtual assistants and chatbots powered by AI technology may engage with patients, answering their queries, delivering individualized health information, and offering support and assistance. This encourages patients to take an active part in managing their health, increasing self-care, and boosting health literacy.

Healthcare practitioners also benefit from AI-driven decision support systems. Clinical decision support systems (CDSS) combine patient data, medical expertise, and AI algorithms to aid healthcare practitioners in making evidence-based choices. By evaluating patient data, treatment guidelines, and medical literature, AI can give suggestions for diagnosis, treatment alternatives, and monitoring plans, helping healthcare practitioners to provide more accurate and efficient care.

However, as AI becomes more incorporated into health care, it presents serious ethical problems. Privacy and security of patient data, fairness and equality in access to AI-enabled treatment, and guaranteeing accountability and transparency in AI algorithms are essential factors that need to be addressed to ensure ethical and responsible usage of AI in patient care.

In conclusion, the influence of AI on patient care delivery is tremendous, transforming all sectors of healthcare. From precise diagnosis to remote monitoring, individualized treatment plans, and patient involvement, AI has the ability to drastically improve patient outcomes, enhance healthcare delivery, and alter the future of healthcare. However, it is vital to manage the ethical and legal difficulties to enable the appropriate and fair integration of AI in patient care.

AI Revolution in Medicine, The Future of Healthcare

6.2 AI in Patient Monitoring and Management

6.2.1 Remote Patient Monitoring and Telehealth

Remote patient monitoring (RPM) and telehealth have developed as essential tools in patient care, particularly in instances when in-person visits are problematic or not practicable. AI plays a key role in boosting the efficacy and efficiency of remote patient monitoring and telehealth services.

Remote patient monitoring includes the use of linked devices and sensors to gather and communicate patient data, such as vital signs, activity levels, and medication adherence, to healthcare practitioners for remote monitoring and analysis. AI systems can assess this constant stream of patient data in real-time, allowing early diagnosis of health worsening and prompt treatments. For example, AI may discover trends or anomalies in vital signs data and provide warnings to healthcare personnel when irregularities are found, allowing timely medical response.

Telehealth refers to the delivery of healthcare services remotely, enabling patients to consult with healthcare experts through video conferencing, phone conversations, or secure messaging systems. AI technologies boost telehealth services by offering intelligent decision assistance and enhancing the accuracy and efficiency of remote consultations. AI-powered chatbots and virtual assistants may support patients by answering their inquiries, delivering symptom evaluations, and offering general health counseling. These systems may triage patients based on their symptoms and send them to suitable treatment pathways, guaranteeing effective use of healthcare resources.

AI algorithms also enable the analysis of telehealth interactions, allowing healthcare practitioners to gather and evaluate valuable information from video consultations or distant examinations. Natural Language Processing (NLP) approaches allow the extraction of significant insights from unstructured data, such as clinical notes and audio transcripts, improving correct documentation and continuity of treatment.

Moreover, AI-driven predictive analytics may enhance remote patient monitoring and telemedicine by identifying high-risk patients who may need extra treatments or closer monitoring. By examining past patient data, AI systems may provide risk ratings and forecast bad occurrences, helping healthcare personnel to proactively intervene and avert probable issues.

However, the implementation of remote patient monitoring and telehealth services comes with hurdles. Ensuring the confidentiality and privacy of patient data during transmission and storage is critical. Healthcare firms must employ effective cybersecurity safeguards and comply to data protection requirements to ensure patient confidentiality.

Furthermore, tackling the digital gap and providing fair access to telehealth services is vital. Adequate infrastructure, consistent internet connection, and availability to essential equipment pose challenges for some patient groups, notably those in underserved places or with poor technology knowledge. Efforts should be made to overcome these gaps and give equitable opportunity for all patients to benefit from AI-enabled remote patient monitoring and telehealth services.

In summary, AI plays a vital role in boosting patient monitoring and management via remote patient monitoring and telehealth. By employing AI algorithms, healthcare practitioners may remotely monitor patients, give timely treatments, and enhance patient outcomes. However, assuring data security, resolving inequities, and safeguarding patient privacy are critical challenges in the adoption and application of AI in remote patient monitoring and telehealth.

6.2.2 Predictive Analytics for Early Disease Detection and Intervention

Predictive analytics driven by AI has the potential to transform early illness identification and intervention, leading to improved patient outcomes and better management of healthcare resources. By analyzing vast amounts of patient data and discovering trends and risk factors, AI algorithms may aid healthcare practitioners in identifying patients at risk of getting certain illnesses at an early stage.

One area where predictive analytics is especially significant is in the identification of persons at risk of chronic illnesses, such as diabetes, cardiovascular disorders, and certain forms of cancer. AI algorithms may scan multiple patient data sources, including electronic health records, genetic information, lifestyle data, and social determinants of health, to identify people who have a greater risk of getting certain disorders. By integrating and evaluating these different data sets, AI models may provide risk ratings or prediction models that assist healthcare practitioners prioritize resources and actions for high-risk people.

In addition to identifying at-risk patients, predictive analytics may also assist in forecasting illness progression and consequences. By examining previous patient data and real-time data streams, AI systems may anticipate disease development trajectories, treatment response, and probable consequences. These predictions allow healthcare practitioners to proactively intervene and tailor treatment strategies, leading to improved patient outcomes.

Moreover, predictive analytics may aid in improving healthcare resource allocation by identifying patients who are likely to gain the most from early treatments. By prioritizing treatments for high-risk patients, healthcare organizations may spend their resources more effectively and efficiently, resulting to improved patient care and decreased healthcare expenditures.

However, the effective adoption of predictive analytics in early illness identification and intervention depends on the availability of high-quality and complete patient data. Data integration from numerous sources, including electronic health records, wearable devices, and genetic databases, is important to offer a comprehensive perspective of patient health. Healthcare institutions must also handle difficulties connected to data privacy, security, and interoperability to guarantee the ethical and appropriate use of patient data.

Furthermore, healthcare practitioners need to examine the influence of AI predictions on patient care and communication. Clear communication techniques should be devised to explain the predictions and their limits to patients, ensuring they understand the aim of therapies and possible results. Maintaining patient trust and resolving issues around data privacy and bias are critical for the effective deployment of predictive analytics in healthcare.

In conclusion, predictive analytics enabled by AI offers enormous promise in early illness identification and intervention. By integrating patient data and AI algorithms, healthcare professionals may identify patients at risk, forecast illness progression, and optimize therapies for improved patient outcomes. However, careful consideration of data quality, privacy, and communication is necessary to enable the ethical and successful use of predictive analytics in healthcare.

6.3 AI in Medication Management and Adherence
6.3.1 Intelligent Medication Dispensing Systems

AI-powered pharmaceutical distribution systems are redefining medication management and adherence, tackling common difficulties such as medication mistakes, non-adherence, and drug-related adverse events. These systems employ AI algorithms and modern technology to enhance the accuracy, safety, and efficiency of pharmaceutical administration operations.

Intelligent medicine distribution systems include numerous characteristics to guarantee the proper drug is provided to the right patient at the right time. They apply machine learning algorithms to properly identify pharmaceuticals based on unique identifiers, such as barcodes or RFID tags, lowering the possibility of prescription mistakes. By cross-referencing patient-specific information, such as electronic health records and medication history, these systems may identify possible drug interactions, allergies, or contraindications, alerting healthcare practitioners and patients to potential hazards.

AI algorithms also allow intelligent pharmaceutical distribution systems to deliver individualized medication instructions and reminders. By examining characteristics such as the patient's age, weight, medical condition, and other pertinent information, these systems may create personalised instructions for medicine delivery, dose modifications, and timing. Automated reminders may be delivered to patients through mobile applications, text messages, or smart devices to increase adherence to drug regimens.

Furthermore, intelligent pharmaceutical distribution systems may track medicine consumption, monitor adherence trends, and send real-time warnings for healthcare practitioners or caregivers when doses are missed or delayed. This allows healthcare practitioners respond swiftly to resolve adherence concerns, give more assistance, or change the treatment plan as required. By increasing drug adherence, these systems help to enhance patient outcomes and minimize healthcare expenditures associated with medication-related problems.

Additionally, AI-powered medicine distribution devices aid to drug inventory management and optimization. By continually monitoring medicine supply levels, these systems may produce automatic notifications when supplies are running low, enabling healthcare institutions maintain appropriate stock and prevent stockouts. This enhances operational efficiency, simplifies drug procurement procedures, and lowers the danger of medication shortages.

However, the adoption of intelligent drug distribution systems involves consideration of many issues. Integration with current electronic health record systems and drug databases is necessary to provide accurate medication identification and patient-specific information retrieval. Data privacy and security safeguards must be in place to secure patient information, since prescription data includes sensitive and personal facts. Furthermore, appropriate training and education for healthcare practitioners, patients, and caregivers are important to enable optimal usage of these systems and optimize their advantages.

In conclusion, AI-driven intelligent medicine distribution systems provide considerable benefits in medication management and adherence. By integrating AI algorithms and modern technology, these solutions promote drug safety, give individualized instructions and reminders, and allow real-time monitoring of prescription adherence. Implementing these systems involves careful consideration of data integration, privacy, security, and user education, but their potential to enhance patient outcomes and expedite pharmaceutical operations is enormous.

6.3.2 Personalized Medication Regimen Optimization

Personalized medicine regimen optimization refers to the use of AI and advanced analytics to design pharmaceutical regimens based on individual patient characteristics, medical history, genetic variables, and other pertinent data. This method tries to optimize the selection of drugs, doses, and treatment regimens to enhance efficacy while reducing possible adverse effects and drug interactions.

AI algorithms examine enormous volumes of patient data, including electronic health records, genetic profiles, clinical guidelines, and research papers, to provide individualized medicine recommendations. These algorithms analyze different characteristics such as age, sex, weight, medical conditions, allergies, comorbidities, medication history, and genetic variants that may effect drug metabolism or reaction.

The AI-driven drug regimen optimization approach contains multiple components. First, the algorithm evaluates the patient's medical history, including diagnoses, test findings, and treatment responses, to discover trends and anticipate how a patient is likely to react to various drugs. This information helps guide the selection of suitable drugs for the patient's condition.

Next, the algorithm evaluates the patient's genetic information, if available, to see whether there are unique genetic differences that may alter medication metabolism or effectiveness. This genetic data may influence judgments on drug selection, dose changes, or alternate therapy alternatives.

The AI system also takes into consideration possible drug-drug interactions, medication adherence trends, and the patient's lifestyle characteristics to improve the prescription regimen. By collecting data from many sources and employing sophisticated analytics methods, the system may detect possible dangers, make dose modifications, or suggest alternative drugs to optimize the overall treatment plan.

Moreover, AI may aid in real-time monitoring and modification of pharmaceutical regimens. For instance, wearable devices or mobile apps linked to the AI system may gather data on a patient's vital signs, symptoms, or side effects, and offer continuous input to healthcare practitioners. This information permits prompt interventions or alterations to the pharmaceutical regimen to guarantee optimum treatment results.

Personalized medicine regimen optimization provides various advantages. It enhances the possibility of picking the most effective drugs for each patient, reducing trial-and-error techniques. By examining individual traits and genetic variables, it helps minimize harmful medication responses and adjust doses based on particular patient requirements. This method may improve treatment results, boost patient satisfaction, and minimize healthcare expenditures associated with inefficient or suboptimal pharmaceutical usage.

However, problems remain in adopting tailored drug regimen optimization. Access to full patient data, including genetic information, is vital but may not always be readily available or easily accessible. Integrating AI algorithms into current electronic health record systems and clinical procedures may be complicated and needs careful preparation. Moreover, guaranteeing data privacy, security, and ethical concerns is vital to secure medical information and sustain patient confidence.

In conclusion, customized medication regimen optimization enabled by AI has the potential to transform medication management by personalizing therapies to specific patients. By utilizing patient-specific data, genetic information, and sophisticated analytics, AI systems may deliver individualized drug recommendations that boost treatment effectiveness, limit unwanted effects, and improve patient outcomes. Overcoming hurdles related to data availability, system integration, and privacy issues will be crucial to fully realize the advantages of individualized medicine regimen optimization in clinical practice.

6.4 AI in Patient Engagement and Education
6.4.1 Virtual Assistants and Chatbots for Patient Support

Virtual assistants and chatbots powered by artificial intelligence (AI) are increasingly being employed in healthcare settings to boost patient engagement and deliver tailored assistance. These intelligent conversational bots may communicate with patients via text or voice-based interfaces, giving support, answering queries, and delivering important health information.

The function of virtual assistants and chatbots in patient assistance is diverse. They may give information on medical issues, treatment alternatives, medication instructions, and lifestyle advice, helping people make educated decisions about their healthcare. Virtual assistants may also give advise on managing chronic diseases, provide reminders for medication adherence or impending visits, and simplify access to healthcare information and services.

AI-powered virtual assistants and chatbots are meant to imitate human-like interactions, employing natural language processing (NLP) algorithms to interpret patient requests and answer correctly. They can decipher complicated medical language and give accurate and relevant information in a user-friendly way. These technologies may also learn and change over time, enhancing their answers depending on user interactions and feedback.

One of the primary benefits of virtual assistants and chatbots is their 24/7 availability, giving patients with rapid access to help and information outside of conventional healthcare settings. Patients may connect with these AI-driven agents via different devices, including as smartphones, tablets, or smart speakers, offering easy and on-demand help.

Virtual assistants and chatbots may be especially effective for patient education. They may give customised instructional information suited to each patient's unique requirements, preferences, and health problems. This tailored approach helps patients better understand their diseases, treatment plans, and self-care practices, allowing them to actively engage in their healthcare journey.

Furthermore, virtual assistants and chatbots may gather and analyze data from patient interactions, yielding significant insights for healthcare practitioners. These insights might assist highlight common patient issues, gaps in understanding, or places where extra support may be necessary. Healthcare businesses may harness this data to improve patient education materials, optimize care delivery systems, and increase overall patient experiences.

However, there are crucial issues when using virtual assistants and chatbots in patient assistance. Ensuring the quality and dependability of the information supplied is vital, since erroneous or misleading replies might have harmful impacts on patient treatment. Maintaining patient privacy and data security is also critical, since these conversational chatbots may handle sensitive health information. Organizations must comply to data protection standards and utilize comprehensive security measures to preserve patient data.

In conclusion, virtual assistants and chatbots powered by AI provide essential assistance and education for patients. By offering individualized and accessible information, they may boost patient involvement, improve health literacy, and enable self-management of health issues. Healthcare practitioners may harness these technologies to expand treatment beyond typical clinical settings, delivering ongoing support and education to patients. However, assuring accuracy, privacy, and security are critical for the effective use of virtual assistants and chatbots in patient assistance.

6.4.2 Personalized Health Information and Education

Personalized health information and education play a significant role in allowing people to make educated choices about their health and well-being. With the breakthroughs in artificial intelligence

(AI), healthcare professionals may harness this technology to give individualized health information and education to patients.

AI can evaluate huge volumes of patient data, including medical records, genetic information, lifestyle variables, and clinical recommendations, to provide personalised health advice. By examining individual attributes, such as age, gender, medical history, and risk factors, AI algorithms may deliver individualized suggestions, preventative actions, and instructional materials.

One use of AI in individualized health information and education is the creation of virtual health assistants. These AI-powered platforms can engage with patients, recognize their individual health requirements, and give personalised instructional information. Virtual health assistants may give explanations of medical issues, treatment alternatives, and self-care practices in a style that is simply understood to each person.

Additionally, AI may enable healthcare professionals provide individualized health education via numerous channels, such as mobile apps, internet platforms, or wearable devices. These technologies may gather real-time data on people' health activities, monitor their progress, and give individualized feedback and recommendations to support their health objectives.

Personalized health information and education may also be targeted to particular demographics or areas. AI algorithms can assess demographic and social factors of health data to discover specific health concerns and generate customized remedies. This method may alleviate health inequities and improve fair access to healthcare services and information.

Furthermore, AI can support continual learning and engagement by changing training materials depending on patient preferences and learning styles. By studying user interactions and comments, AI systems may adjust the delivery of information to boost comprehension and engagement.

However, ethical issues must be taken into account when offering individualized health information and education utilizing AI. Patient privacy and data security should be respected, and patients should have authority over the gathering and use of their health data. Transparency in how AI algorithms function and the sources of information utilized is also vital to retain confidence and assure the accuracy and dependability of the individualized suggestions supplied.

In conclusion, AI allows the dissemination of individualized health information and education, allowing people to take an active part in controlling their health. By integrating patient data and AI algorithms, healthcare professionals may give personalised advice, instructional materials, and treatments to suit people' particular health requirements. This individualized strategy may enhance health outcomes, promote health literacy, and encourage patient participation and empowerment. However, ethical issues about privacy, security, and openness should drive the deployment of AI in providing individualized health information and education.

6.5 AI in Mental Health and Well-being
6.5.1 AI-Based Mental Health Screening and Assessment

Mental health is a significant part of total well-being, and early diagnosis and evaluation of mental health disorders are important for prompt intervention and treatment. Artificial intelligence (AI) has emerged as a potential tool in the area of mental health by giving unique ways for screening and diagnosing mental health disorders.

AI-based mental health screening includes the use of machine learning algorithms to scan multiple data sources, such as self-reported surveys, social media postings, electronic health records, and physiological data, to detect possible symptoms of mental health issues. These algorithms may uncover patterns, correlations, and risk variables that may suggest the existence of mental health disorders.

One significant benefit of AI-based mental health screening is its capacity to process vast volumes of data rapidly and effectively. By analyzing enormous information, AI computers may give more accurate and impartial judgments, eliminating the possibility for human biases. Furthermore, AI screening techniques may be readily scaled, providing for larger reach and accessibility in identifying people who may benefit from further examination or intervention.

AI-based evaluation tools may potentially assist in identifying and monitoring mental health issues. These technologies may evaluate speech patterns, facial expressions, and other behavioral clues to give insights about an individual's mental well-being. For example, natural language processing methods may be applied to examine the content and mood of spoken or written material to determine an individual's emotional state.

Moreover, AI may promote the creation of virtual mental health assistants or chatbots that can participate in discussions with people to give support, information, and resources. These virtual assistants may apply natural language processing and machine learning algorithms to analyze and react to people' worries, give psychoeducation, and suggest coping techniques.

While AI-based mental health screening and evaluation provide considerable promise, some ethical problems must be addressed. Privacy and confidentiality of sensitive mental health data should be guaranteed, and people should have discretion over the use and sharing of their information. Additionally, there is a need for continual validation and modification of AI algorithms to increase their accuracy and dependability. Transparency in how AI algorithms function and the limits of their judgments is vital to retain confidence between patients and healthcare practitioners.

In conclusion, AI-based mental health screening and evaluation have the potential to change mental healthcare by allowing early identification and intervention. Through the examination of diverse data sources and the use of machine learning algorithms, AI can offer objective and scalable tools for spotting probable mental health disorders. However, ethical issues surrounding privacy, data security, and algorithm openness should govern the development and deployment of AI-based mental health screening and evaluation technologies.

6.6 AI-Enabled Precision Rehabilitation
6.6.1 Personalized Rehabilitation Planning and Monitoring

Rehabilitation serves a key role in helping patients recover from accidents, surgeries, or chronic diseases. AI technology has the potential to improve rehabilitation by providing tailored planning and monitoring of rehabilitation programs.

AI algorithms may assess diverse patient-specific data, such as medical records, physiological measures, and mobility data, to build individualized rehabilitation regimens. These algorithms can find patterns and connections in the data to predict the most effective workouts, treatments, and interventions for each person. By examining criteria including the patient's health, skills, and objectives, AI can modify the rehabilitation strategy to maximize efficacy and improve results.

Moreover, AI can assist real-time monitoring of patients throughout their recovery process. Wearable equipment, such as sensors or smart clothes, may gather movement and physiological data, which can be continually evaluated by AI algorithms. This enables for instant feedback and customization of the rehabilitation program depending on the patient's development and performance.

AI may identify deviations from the required movement patterns or physiological characteristics and give coaching or warnings to guarantee appropriate form and minimize any dangers or accidents.

Another component of AI-enabled precision rehabilitation is the integration of virtual reality (VR) and augmented reality (AR) technology. These immersive technologies may create dynamic and engaging rehabilitation settings, imitating real-world events or offering visual signals and direction for activities. AI algorithms may track the patient's motions inside the VR or AR environment and give real-time feedback and coaching, boosting the efficacy of rehabilitation sessions.

AI-enabled precision rehabilitation provides various advantages, including tailored treatment regimens, continuous monitoring, and quick feedback. By adapting rehabilitation programs to the individual requirements and skills of each patient, AI can enhance results and increase rehabilitation efficiency. Furthermore, the real-time monitoring and feedback given by AI technology may boost patient involvement and motivation, leading to greater adherence and development.

However, ethical issues need to be addressed when applying AI-enabled precision rehabilitation. Patient privacy and data security should be respected, and informed permission should be acquired about the use of AI technology in rehabilitation. Additionally, healthcare practitioners should retain an active involvement in reviewing and interpreting the AI-generated suggestions to ensure they correspond with the patient's particular requirements and preferences.

In summary, AI-enabled precision rehabilitation shows considerable potential in customizing and optimizing rehabilitation regimens. By integrating patient-specific data and real-time monitoring, AI algorithms may generate personalised treatment regimens and give ongoing advice and feedback. Ethical concerns should drive the use of AI in rehabilitation to preserve patient privacy and guarantee the engagement of healthcare professionals in decision-making.

6.6.2 AI-Assisted Prosthetics and Assistive Devices

Advancements in artificial intelligence (AI) technology have brought up new possibilities in the realm of prosthetics and assistive technologies. AI may be used to increase the functioning, usefulness, and flexibility of prosthetic limbs and other assistive devices, enhancing the quality of life for those with limb loss or disability.

AI algorithms may be applied to construct intuitive and intelligent control systems for prosthetic limbs. By evaluating electromyography (EMG) signals or other physiological inputs, AI can comprehend the user's intended motions and transform them into equivalent actions of the prosthetic limb. This offers more natural and accurate control, enabling users to complete difficult activities and motions with more comfort.

Furthermore, AI can enable the construction of prosthetic limbs with adaptive capacities. Machine learning algorithms can continually learn from the user's actions and preferences, enabling the prosthetic limb to adapt and modify its behavior appropriately. This tailored adaptation increases the comfort, functionality, and efficiency of the prosthetic limb, boosting the user's entire experience and performance.

In addition to prosthetic limbs, AI technology may augment other assistive devices used for movement, communication, and everyday living chores. AI-powered wheelchairs may use computer vision and sensor technology to explore areas, avoid obstacles, and give intelligent support to users. AI algorithms may also provide speech recognition and natural language processing, enabling those with communication difficulties to communicate with assistive devices via voice commands.

One of the key benefits of AI-assisted prostheses and assistive devices is their capacity to learn and adapt over time. Machine learning algorithms may assess user data, such as movement patterns, user preferences, and environmental context, to constantly enhance device performance and personalization. This adaptive learning feature guarantees that the gadgets change and match the changing demands of the user.

Ethical issues are critical when incorporating AI into prostheses and assistive technologies. Privacy and security of user data should be protected, and informed permission should be acquired about the collection and use of personal information. Transparency and explainability of AI algorithms are also vital to develop confidence between people and their AI-assisted gadgets. Additionally, equal access to AI-enabled prostheses and assistive devices should be addressed to guarantee that those with disabilities may benefit from these technological developments.

In conclusion, AI-assisted prosthetics and assistive technologies have the potential to transform the area of rehabilitation and enhance the lives of those with limb loss or disability. By integrating AI algorithms for control, adaptability, and learning, these gadgets may deliver more intuitive and tailored help to consumers. Ethical concerns should govern the development and implementation of AI technologies in this sector to guarantee privacy, transparency, and fair access for all those in need.

6.7 AI in End-of-Life Care and Palliative Care
6.7.1 AI-Driven Symptom Management and Comfort Care

AI has the potential to play a big role in improving end-of-life care and palliative care by boosting symptom management and delivering comfort to patients. By integrating AI technology, healthcare practitioners may better recognize and manage the particular requirements and preferences of patients reaching the end of life.

One significant use of AI in end-of-life care is the treatment of symptoms reported by patients. AI algorithms may evaluate patient data, including medical records, vital signs, and patient-reported symptoms, to detect patterns and connections. This information may be used to construct prediction models that assist healthcare practitioners anticipate and treat symptoms more efficiently. For example, AI may assist determine the probability and severity of pain, nausea, or dyspnea based on multiple parameters such as disease progression, treatment history, and patient characteristics. This allows healthcare providers to proactively intervene and give timely therapies to lessen uncomfortable symptoms.

AI may also aid in the customization of end-of-life care by addressing unique patient preferences and objectives. Machine learning algorithms may assess patient data and patient-reported outcomes to identify preferences on pain management, treatment alternatives, and quality-of-life objectives. This information may be utilized to adapt the treatment plan to each patient, ensuring that therapies correspond with their beliefs and objectives. For example, AI may assist assess whether a patient chooses aggressive treatment methods or prioritizes comfort care and support in their dying days.

Another key part of end-of-life care is encouraging emotional well-being and fostering communication. AI-powered chatbots or virtual assistants may give a forum for patients and their families to voice their worries, anxieties, and emotions. These AI systems may employ natural language processing to interpret and reply to patient enquiries, give sympathetic assistance, and provide resources for emotional and psychological care. AI may also aid in enabling tough talks regarding end-of-life decisions, helping patients and their families make educated choices and ensuring their wishes are fulfilled.

Ethical issues are crucial when adopting AI in end-of-life care and palliative care. Patient privacy and confidentiality must be respected, and permission for data use should be acquired. It is vital to retain human connection and guarantee that AI technology does not replace compassionate and empathic human relationships. AI should be considered as a tool to improve care delivery rather than a substitute for the human touch and presence that is crucial in end-of-life care.

In summary, AI-driven symptom management and comfort care have the potential to dramatically enhance the quality of end-of-life care and palliative care. By employing AI algorithms to anticipate and manage symptoms, tailor care plans, and improve communication, healthcare practitioners may better serve patients and their families throughout this vital time of life. Ethical concerns should lead the appropriate and compassionate integration of AI technology to guarantee that patient autonomy, privacy, and emotional well-being are valued.

6.7.2 Predictive Analytics for Palliative Care Planning

Predictive analytics, driven by artificial intelligence (AI), may play a crucial role in palliative care planning by offering insights and predictions that enable healthcare practitioners anticipate and manage the shifting needs of patients. By evaluating vast amounts of patient data and discovering patterns and correlations, predictive analytics may enhance proactive decision-making and improve the quality of treatment for patients in need of palliative care.

One significant use of predictive analytics in palliative care planning is the identification of patients who may benefit from palliative care treatments. AI algorithms may examine patient data, including medical records, clinical signs, and demographic information, to identify people who are at greater risk of suffering major symptoms, disease progression, or worsening in their health. This proactive identification helps healthcare personnel to commence talks and actions early, ensuring that patients get timely palliative care assistance.

Predictive analytics may also aid in forecasting and controlling symptom load in palliative care patients. By evaluating past patient data and combining real-time data, AI algorithms may develop prediction models that assess the frequency and severity of symptoms such as pain, dyspnea, or nausea. These models may take into consideration many parameters, including illness development, treatment history, and patient characteristics, to generate tailored forecasts. This information helps healthcare professionals to design proactive symptom management programs, ensuring that patients' symptoms are adequately handled and their comfort is maximized.

Furthermore, predictive analytics may aid in projecting healthcare resource usage and directing resource allocation in palliative care settings. By assessing patient data and considering criteria such as illness trajectory, comorbidities, and historical healthcare use trends, AI algorithms can estimate the degree of care and services that patients may need. This information may aid healthcare practitioners in assigning appropriate resources, including personnel, equipment, and pharmaceutical supplies, to ensure that patients get the necessary degree of support and that resources are effectively used.

Ethical issues are critical when employing predictive analytics in palliative care planning. It is necessary to maintain patient privacy and confidentiality, as well as gain informed permission for data use. Transparency in the prediction models and algorithms is vital to develop confidence and allow collaborative decision-making between healthcare practitioners, patients, and their families. Human judgment and compassionate care should always be at the forefront, and predictive analytics should be considered as a supporting tool to assist clinical decision-making rather than replacing the clinician's knowledge and patient-centered care.

In summary, predictive analytics may considerably increase palliative care planning by identifying patients who may benefit from palliative care treatments, forecasting symptom load, and improving resource allocation. By employing AI technology to evaluate patient data and build predictive models, healthcare practitioners can anticipate and meet the increasing requirements of palliative care patients, providing prompt and individualized treatment. Ethical issues and the humanistic approach to palliative care should lead the incorporation of predictive analytics to protect patient autonomy, privacy, and the provision of compassionate end-of-life care.

CHAPTER 7: ETHICAL AND LEGAL CONSIDERATIONS

7.1 Introduction to Ethical and Legal Considerations in AI

7.1.1 Importance of Ethical and Legal Frameworks in AI Development

1. The fast growth of artificial intelligence (AI) technology has prompted important ethical and legal problems. It is vital to construct rigorous frameworks that govern the development, implementation, and usage of AI systems.

2. Ethical concerns in AI guarantee that the technology corresponds with social norms, protects human rights, and promotes justice and responsibility.

3. Legal frameworks give a legislative framework to control AI systems, covering concerns such as responsibility, privacy, and data protection.

4. Ethical and legal frameworks in AI development are crucial to create public trust, safeguard people' rights, and enable responsible and productive use of AI technology.

5. Ethical and legal issues in AI assist limit possible risks and disadvantages connected with biased algorithms, privacy breaches, and lack of transparency.

6. These frameworks also enhance openness, explainability, and accountability in AI systems, allowing for greater understanding and assessment of its choices and behaviors.

7. Ethical and legal issues are important to address the possible effect of AI on employment, social inequities, and human autonomy.

AI Revolution in Medicine, The Future of Healthcare

8. By developing ethical and legal frameworks, stakeholders may work together to guarantee that AI is created and implemented in a way that benefits people, companies, and society as a whole.

Overall, ethical and legal frameworks play a significant role in directing the development and deployment of AI systems, ensuring they comply to society norms, safeguard people' rights, and encourage responsible and constructive use of AI technology.

7.2 Ethical Principles in AI Development and Deployment

7.2.1 Fairness and Bias Mitigation

1. Fairness is a basic ethical ideal in AI that attempts to promote fair treatment and opportunity for all humans, regardless of their qualities or origins.

2. Bias in AI systems may originate from biased training data, poor model design, or biased decision-making procedures. It may lead to biased results and perpetuate social disparities.

3. Mitigating prejudice in AI entails recognizing and correcting possible biases in data, algorithms, and decision-making processes. This involves assuring representative and varied training datasets, analyzing and fixing algorithmic biases, and encouraging openness in the decision-making process.

4. Fairness metrics, such as demographic parity, equal opportunity, and predictive parity, may be applied to analyze and prevent prejudice in AI systems.

5. Ethical principles and standards, such as the use of impartial data gathering and validation methodologies, may assist developers and organizations proactively address fairness and eliminate prejudice in AI systems.

6. Ongoing monitoring and assessment of AI systems are important to discover and remedy any developing biases and assure continual development in fairness.

7. Collaboration amongst multidisciplinary teams, comprising data scientists, ethicists, and domain specialists, may assist include multiple viewpoints and assure the creation of fair and impartial AI systems.

8. Ethical issues relating to fairness and prejudice should also examine the possible influence of AI on disadvantaged communities and prevent any potential damage or discrimination.

In summary, fairness and bias reduction are key ethical considerations in AI research and deployment. By eliminating biases and supporting equal treatment, AI systems may be developed to contribute to a more just and inclusive society.

7.2.2 Transparency and Explainability

• Transparency and explainability in AI systems relate to the ability to comprehend and offer explicit explanations for the choices and actions done by AI algorithms.

• Transparent AI systems allow insight into their inner workings, including the data utilised, the algorithms applied, and the decision-making processes.

• Explainable AI (XAI) strives to give human-understandable explanations for the results and predictions made by AI systems.

• Transparency and explainability are crucial ethical standards since they foster accountability, trust, and human supervision in AI decision-making.

• Transparent AI systems allow stakeholders, including end-users, regulators, and policymakers, to understand and assess the behavior and effect of AI algorithms.

- Explainability in AI is especially vital in key sectors such as healthcare, where the ability to comprehend and defend the judgments made by AI systems is crucial.

- Techniques for boosting transparency and explainability in AI include model interpretability approaches, such as feature significance analysis, rule extraction, and surrogate models.

- Standardization and standards for openness and explainability may assist promote best practices and guarantee uniform methods across various AI applications and sectors.

- Balancing openness and explainability with the need to preserve sensitive information and retain proprietary knowledge is a problem that needs careful thought.

- Ethical issues relating to openness and explainability also entail addressing the possible hazards of manipulation, fraud, or unintended effects that may occur from offering explanations.

- Ongoing research and cooperation among academics, practitioners, and policymakers are crucial to progress the development and implementation of transparent and explainable AI systems.

In conclusion, openness and explainability are crucial ethical criteria in AI research and deployment. By offering insights into AI decision-making, these principles create trust, accountability, and the capacity to detect and eliminate possible biases or mistakes in AI systems.

7.2.3 Privacy and Data Protection

1. Privacy and data protection are essential ethical and legal issues in AI research and implementation.

2. AI systems generally depend on vast volumes of personal data for training, testing, and generating predictions. Protecting the

privacy of people and guaranteeing the secure management of their data is crucial.

3. Privacy refers to the right of people to regulate the acquisition, use, and dissemination of their personal information.

4. Data protection entails establishing procedures to secure personal data from unlawful access, use, or disclosure.

5. Ethical considerations such as data reduction, purpose restriction, and informed permission play a significant role in respecting privacy and data protection.

6. Data anonymization and de-identification methods may be applied to eliminate or decrease personally identifying information from datasets used in AI research.

7. Privacy-enhancing technologies (PETs) such as differential privacy, secure multi-party computing, and federated learning may be utilized to ensure data privacy while allowing for collaborative AI model building.

8. Compliance with current privacy rules and regulations, such as the General Data Protection Regulation (GDPR), is vital when designing and implementing AI systems that handle personal data.

9. Organizations should develop explicit data governance rules, including data access limits, data preservation standards, and protocols for resolving data breaches or incidents.

10. User permission and openness regarding data collection and use procedures are crucial parts of ensuring privacy in AI systems.

11. Regular privacy impact assessments and audits may assist identify and manage possible privacy concerns and assure continuing compliance with privacy legislation.

12. Collaboration between AI engineers, legal experts, and privacy specialists is important to traverse the complicated terrain of privacy and data protection in AI.

13. Privacy and data protection issues should be addressed throughout the full lifetime of AI systems, from data gathering and model training through deployment and continuous monitoring.

In summary, privacy and data protection are crucial ethical and legal problems in AI. Respecting people' privacy rights, adopting adequate data protection measures, and complying with applicable laws and regulations are vital for establishing trustworthy and responsible AI systems.

7.2.4 Accountability and Responsibility

1. Accountability and accountability are crucial ethical considerations in AI development and deployment.

2. Accountability refers to the requirement of people and organizations to accept responsibility for the acts and choices made by AI systems.

3. Responsibility requires assessing the possible implications of AI systems on persons, society, and the environment and adopting necessary efforts to avoid risks and achieve constructive results.

4. AI developers and organizations should be responsible for the design, development, and use of AI systems, including the effect on persons' rights, well-being, and autonomy.

5. Transparent and responsible AI governance frameworks should be built to guarantee responsibility throughout the AI lifecycle.

6. Clear lines of accountability and decision-making power should be developed to clarify who is responsible for AI-related choices and activities.

AI Revolution in Medicine, The Future of Healthcare

7. It is crucial to build processes for resolving biases, errors, and unexpected effects that may come from AI systems, taking remedial steps, and learning from mistakes.

8. The notion of algorithmic accountability stresses the necessity for openness, explainability, and auditing of AI systems to hold them responsible for their choices and actions.

9. Stakeholder engagement and participation, including feedback from impacted persons and groups, may assist ensure that AI systems conform with society values and solve issues linked to responsibility.

10. Collaboration between AI developers, legislators, regulatory authorities, and other stakeholders is vital to define rules, standards, and procedures for accountability in AI.

11. Ethical frameworks and codes of conduct, such as the IEEE Ethically Aligned Design, may give direction on fostering accountability and responsibility in AI development.

12. Legal frameworks and laws may also play a role in establishing the legal obligations and liabilities of AI developers and deployers.

13. Continuous monitoring, review, and auditing of AI systems are important to evaluate their effectiveness, detect any downsides, and resolve any ethical or legal problems that may develop.

14. Ultimately, accountability and responsibility in AI strive to guarantee that the advantages of AI are maximized while avoiding any possible negative repercussions and defending the well-being and rights of people and society as a whole.

7.3 Ethical Decision-Making in AI Systems

7.3.1 Ethical Design and Development Practices

1. Ethical design and development processes are vital for ensuring that AI systems accord with ethical ideals and values.

2. AI developers should take a proactive approach to ethics, including ethical issues across the full development lifespan of AI systems.

3. Ethical design entails addressing the possible consequences of AI systems on numerous stakeholders, including people, communities, and society as a whole.

4. Developers should emphasize the well-being, safety, and autonomy of humans impacted by AI systems and seek to prevent injury, prejudice, and injustice.

5. The development process should involve diverse and multidisciplinary teams to include many viewpoints and avoid prejudices and blind spots.

6. Ethical principles and frameworks, such as the IEEE Global Initiative on Ethics of Autonomous and Intelligent Systems, may give significant direction for ethical design and development methods.

7. Transparency and explainability are crucial components of ethical AI systems. Developers should aim to make the decision-making processes of AI systems intelligible and give explanations for the results they create.

8. Data quality and integrity should be protected to prevent propagating biases and injustice. Bias identification and mitigation strategies should be utilized throughout the data collection, preprocessing, and model training phases.

9. Ethical concerns should be included into the selection and use of training data to prevent the continuation of discriminating or harmful tendencies.

10. Regular testing, validation, and evaluation of AI systems should be done to examine their performance, dependability, and adherence to ethical norms.

11. User input and public interaction may give vital insights into the ethical implications of AI systems and help guide their design and development.

12. Collaboration with ethicists, social scientists, and other relevant professionals may promote the identification of ethical dilemmas and the creation of acceptable solutions.

13. Ethical review boards or committees may be created inside companies to analyze the ethical implications of AI initiatives and assure compliance with ethical principles and legislation.

14. Continuous monitoring and post-deployment evaluation of AI systems are important to handle developing ethical problems and adapt to changing society expectations.

15. Regular ethics training and instruction for AI developers and practitioners may enhance awareness and comprehension of ethical concepts and assist them make ethical judgments throughout the AI development process.

16. Ultimately, ethical design and development approaches strive to produce responsible and trustworthy AI systems that emphasize ethical issues and contribute positively to society.

7.3.2 Ethical Use and Impact Assessment

1. Ethical usage and effect evaluation entails examining the possible ethical implications and social repercussions of AI systems before, during, and after their deployment.

2. It strives to identify and resolve ethical issues, assure responsible AI usage, and limit possible downsides to persons and society.

3. Ethical usage evaluation entails examining the intended and unintentional outcomes of AI systems, as well as their conformity with ethical ideals and regulatory constraints.

4. Impact evaluation comprises examining the possible social, economic, cultural, and environmental repercussions of AI systems to ensure that their deployment does not result in negative outcomes or worsen existing imbalances.

5. Key issues in ethical usage and effect assessment include justice, openness, privacy, responsibility, and the well-being and rights of persons.

6. Fairness evaluation entails investigating whether AI systems display prejudice, discrimination, or unjust results across various demographic groups. Techniques such as fairness measurements and fairness-aware learning may be used to identify and counteract prejudice.

7. Transparency assessment focuses on understanding the decision-making processes and underlying algorithms of AI systems. It entails analyzing whether AI systems are explainable and interpretable, allowing people to comprehend how choices are obtained.

8. Privacy evaluation entails examining the data collecting, storage, and processing procedures of AI systems to guarantee compliance with privacy standards and safeguard the confidentiality of personal information.

9. Accountability evaluation investigates the processes in place to assign accountability for the acts and choices of AI systems. It involves evaluating questions of culpability, redress, and the capacity to rectify possible damages created by AI systems.

10. Ethical usage and effect evaluations should incorporate a multidisciplinary approach, involving input from ethicists, subject specialists, impacted persons or groups, and other stakeholders.

11. Guidelines and guidelines, such as the Ethical Guidelines for Trustworthy AI created by the European Commission, may give a formal way to undertaking ethical usage and impact evaluations.

12. Continuous monitoring and assessment of AI systems' impacts and input from users and stakeholders are necessary to detect and solve any developing ethical dilemmas or unexpected repercussions.

13. Ethical usage and effect evaluation should be a continuous process throughout the lifespan of AI systems, as technology improves and social values and norms change.

14. Findings and suggestions from ethical usage and impact evaluations should guide the creation of rules, legislation, and best practices for responsible AI deployment.

15. Collaboration and knowledge-sharing across enterprises, researchers, politicians, and the public are vital to enable full ethical usage and effect evaluations and develop ethical AI practices.

16. Ultimately, ethical usage and effect assessment encourages the responsible and accountable deployment of AI systems, helping to create confidence and guarantee that AI technologies contribute constructively to society.

7.3.3 Ethical Governance and Oversight

1. Ethical governance and oversight refers to the procedures, frameworks, and processes in place to assure the ethical development, deployment, and usage of AI systems.

2. It entails developing principles, rules, and standards that regulate the ethical behaviour of companies and people engaged in AI research and implementation.

3. Ethical governance attempts to encourage responsible AI activities, address ethical problems, and assure compliance with legal and regulatory obligations.

4. Key components of ethical governance and monitoring include:

1. Ethical frameworks: Organizations should establish and adhere to ethical frameworks that give advice on responsible AI development and implementation. These frameworks may contain concepts such as justice, openness, accountability, and privacy.

2. Ethical review boards: Organizations may form ethical review boards or committees to analyze and assess the ethical aspects of AI research. These boards may comprise specialists in ethics, law, and related fields to give independent and objective judgements.

3. Ethical guidelines and codes of conduct: Organizations should design and convey explicit rules and codes of conduct that describe ethical expectations for personnel and stakeholders participating in AI development and deployment. These rules may encompass topics such as bias mitigation, data privacy, and openness.

4. Ethical training and awareness: Organizations should conduct training and awareness initiatives to educate personnel about ethical issues in AI. This might involve fostering knowledge of ethical ideas, enhancing awareness of possible biases and hazards, and supporting ethical decision-making.

5. Ethical risk assessments: Organizations should undertake frequent ethical risk assessments to identify and solve any ethical hazards and issues connected with AI systems. This might entail examining the social, cultural, and environmental consequences of

AI deployments and assessing the possible dangers to persons and communities.

6. External supervision and regulation: Regulatory authorities and governmental organizations may play a role in the ethical governance and monitoring of AI. They may set rules, norms, and regulations to guarantee ethical conduct and defend the rights and interests of persons.

7. Transparency and accountability mechanisms: Ethical governance should foster transparency and accountability in AI systems. This might entail methods like as documentation and record-keeping of AI algorithms and data sources, external audits or reviews, and systems for resolving complaints or concerns from impacted persons.

8. Continuous review and improvement: Ethical governance should be an iterative process, with continuous evaluation and improvement of policies and procedures. This might entail monitoring the performance and effect of AI systems, obtaining input from stakeholders, and changing ethical rules and frameworks as appropriate.

9. Collaboration and engagement: Ethical governance should entail collaboration and involvement with stakeholders, including people impacted by AI systems, advocacy organizations, specialists in ethics and AI, and the wider public. This may assist guarantee that varied viewpoints and opinions are considered in the decision-making process.

Ethical governance and supervision are vital for sustaining public confidence, defending individual rights, and maximizing the advantages of AI while reducing the threats. By setting explicit ethical standards and fostering responsible behaviors, companies may guarantee that AI systems are created and deployed in a way that corresponds with society values and ethical principles.

AI Revolution in Medicine, The Future of Healthcare

7.4 Legal Frameworks for AI Regulation

7.4.1 Intellectual Property Rights

1. Intellectual property rights (IPR) relate to legal rights given to people or organizations for their innovations or inventions. In the context of AI, intellectual property plays a key role in preserving the rights and interests of AI developers and stakeholders.

2. AI systems may provide valuable outputs, such as algorithms, models, datasets, and software, which may be eligible for different types of intellectual property protection. Some major factors connected to intellectual property rights in AI include:

1. Patents: Patents protect inventions or developments that are unique, non-obvious, and have industrial use. In the area of AI, patent protection may apply to innovative algorithms, methodologies, or AI applications. Obtaining a patent gives the inventor with exclusive rights to the innovation for a certain duration, enabling them to restrict anyone from using, producing, or selling the patented technology without license.

2. Copyright: Copyright protects original creative works, such as software code, databases, and AI models. It guarantees the author exclusive rights over the reproduction, dissemination, and modification of their work. Copyright protection emerges immediately with the production of the work, without the necessity for registration. AI developers should be aware of copyright laws to ensure that they respect the intellectual property rights of others and get relevant permits for utilizing copyrighted information.

3. Trade secrets: Trade secrets refer to sensitive and valuable knowledge that offers a competitive advantage to a corporation. In the context of AI, trade secrets might include proprietary algorithms, training procedures, or unique datasets. Protecting trade secrets entails establishing procedures to protect their confidentiality and limiting access to authorized personnel. Legal frameworks, like as

non-disclosure agreements (NDAs), may assist safeguard trade secrets.

4. Data rights: With the growing dependence on data for AI development, problems emerge around data ownership and rights. Intellectual property laws normally do not cover raw data, since they mainly protect creative or original manifestations. However, companies may construct data use agreements, licenses, or contracts to regulate the gathering, sharing, and consumption of data.

5. Open-source licensing: Open-source software licenses permit the distribution, modification, and exchange of software code. Many AI frameworks and tools are accessible under open-source licenses, enabling developers to access, alter, and contribute to the development of AI technology. Open-source licensing may enable cooperation and innovation in the AI field.

6. Licensing agreements: Licensing agreements play a role in commercializing AI technology. These agreements specify the terms and circumstances under which AI technology, algorithms, or models may be utilized, disseminated, or incorporated into other systems. Licensing agreements may incorporate licensing fees, use limits, and obligations for intellectual property ownership.

It's crucial for AI developers and organizations to understand and manage the intellectual property environment to protect their own discoveries and respect the intellectual property rights of others. Seeking legal guidance and engaging intellectual property specialists may assist assure compliance with relevant laws and maximize the advantages of AI while protecting intellectual property rights.

7.4.2 Liability and Accountability

In the context of AI, liability and accountability relate to the legal and ethical responsibility for the acts and choices made by AI systems. As AI gets increasingly integrated into numerous fields, assessing culpability and maintaining accountability becomes critical. Here are some major elements connected to responsibility and accountability in AI:

1. Traditional liability frameworks: Existing legal liability frameworks are frequently founded on notions of negligence, strict responsibility, or product liability. However, applying these frameworks to AI systems may be complicated owing to the independent and self-learning nature of AI. Questions emerge around who should be held accountable when an AI system causes damage or makes mistakes.

2. Product responsibility: AI developers and manufacturers may be liable to product liability laws when their AI systems are regarded faulty and cause damage to users or others. Product responsibility involves establishing that the AI system had a fault, the deficiency caused injury, and the system was being utilized as intended or reasonably anticipated.

3. Operator liability: Operators or users of AI systems may also face responsibility for the activities and results of AI systems under their control. If an operator fails to properly monitor or maintain the AI system, resulting in injury or mistakes, they may be held accountable.

4. Algorithmic transparency: Transparency of AI algorithms is vital for creating accountability. It entails making the decision-making processes of AI systems intelligible and interpretable by humans. Transparent algorithms enable users, regulators, and stakeholders to examine the fairness, biases, and possible hazards connected with the system's results.

5. Ethical concerns: While legal responsibility focuses on the adherence to legal responsibilities, ethical considerations go beyond legal requirements. Ethical frameworks support the appropriate

development and deployment of AI systems, ensuring that choices are made with justice, transparency, and the best interests of people and society in mind. Adhering to ethical norms may help limit possible damage and develop confidence in AI systems.

6. Regulatory frameworks: Governments and regulatory agencies are actively working on building new rules and regulations to handle the difficulties brought by AI. These frameworks strive to create clear norms and obligations surrounding responsibility and accountability. They may include rules for risk assessment, impact assessments, safety standards, and reporting procedures to enable responsible AI development and implementation.

7. Human monitoring and control: Maintaining human oversight and control over AI systems might help apportion responsibility and accountability. Human engagement in decision-making processes, especially in crucial or high-stakes areas, may limit risks and guarantee that AI systems function within legal and ethical constraints.

Establishing responsibility and accountability in the context of AI is a continuous task, requiring coordination among policymakers, legal experts, AI developers, and other stakeholders. Striking a balance between innovation and preserving the rights and well-being of persons is vital to build trust and ethical usage of AI technology.

7.4.3 Data Governance and Ownership

Data governance and ownership are crucial parts of AI legislation and legal systems. With the rising dependence on data for AI systems, it is vital to address the concerns of data governance and ownership. Here are major concerns linked to data governance and ownership in the context of AI:

1. Data ownership: Determining who owns the data used in AI systems may be challenging. Data ownership may rely on several aspects, such as legal agreements, data gathering procedures, permission, and relevant legislation. In certain circumstances, data ownership may be shared between the persons giving the data, the organizations collecting the data, and the individuals or entities creating insights from the data using AI.

2. Data protection laws: Data protection rules and regulations, such as the General Data Protection Regulation (GDPR), control the acquisition, use, and sharing of personal data. These laws frequently provide people particular rights and control over their personal data, including the ability to view, modify, and delete their data. AI systems must comply with appropriate data protection legislation to safeguard the privacy and security of people' data.

3. Data governance frameworks: Establishing effective data governance frameworks is critical for safe AI usage. Data governance comprises rules, methods, and practices for managing data throughout its lifespan, including data collection, storage, processing, sharing, and disposal. These frameworks guarantee that data is managed ethically, safely, and in conformity with current laws and regulations.

4. permission and data usage: Obtaining informed and explicit permission from persons for data collection and use is an essential ethical topic. AI systems should adhere to the concepts of purpose restriction and data minimization, ensuring that data is utilized only for the intended purposes and not held longer than required.

5. Data anonymization and de-identification: Anonymizing or de-identifying data may assist safeguard individual privacy. AI systems should contain strategies to erase or obscure personally identifying information (PII) from data, lowering the danger of re-identification.

6. Data sharing and collaboration: AI systems may need access to big and varied datasets for training and validation. Data sharing agreements and standards should be designed to promote responsible data exchange while protecting privacy and confidentiality. Collaboration between data owners, researchers, and organizations may lead to more robust and accurate AI models.

7. Accountability and transparency: Clear accountability procedures should be in place to monitor and trace the usage of data in AI systems. Organizations should be honest about their data collecting and use methods, telling people about how their data is being used and for what goals.

8. International considerations: Data governance and ownership may differ between countries. Organizations operating in various countries must comply with the data protection and privacy legislation of each location. International cooperation and harmonization of data governance systems might assist overcome these difficulties.

Developing thorough data governance frameworks, honoring data ownership rights, and complying with data protection rules are vital for creating confidence in AI systems. These safeguards preserve people' privacy, assure responsible data use, and encourage ethical and legal practices in the AI ecosystem.

7.4.4 Privacy and Data Protection Laws

Privacy and data protection regulations play a critical role in controlling the acquisition, use, storage, and sharing of personal data in the context of AI. As AI systems typically depend on enormous volumes of data, including personal information, it is vital to comply to privacy and data protection rules to ensure people' private rights. Here are major components of privacy and data protection regulations applicable to AI:

AI Revolution in Medicine, The Future of Healthcare

1. General Data Protection Regulation (GDPR): The GDPR, established in the European Union, defines stringent guidelines for the protection of personal data. It defines principles, such as lawfulness, fairness, and openness in data processing, and provides people with rights, including the ability to access, correct, and delete their personal data. AI systems working inside the EU must comply with the GDPR's obligations.

2. Health Insurance Portability and Accountability Act (HIPAA): HIPAA is a U.S. statute that guarantees the privacy and security of people' protected health information (PHI). It applies to healthcare providers, health plans, and healthcare clearinghouses. When AI systems handle or retain PHI, compliance with HIPAA is critical to preserve patient privacy and ensure data security.

3. California Consumer Privacy Act (CCPA): The CCPA is a state-level privacy legislation in California, United States, that offers consumers some rights over their personal data. It mandates firms to declare their data collecting and use policies and lets customers to opt-out of the selling of their personal information. AI systems that handle personal data of California citizens must comply with CCPA regulations.

4. Other national and regional data protection laws: Many nations have their own data protection laws that control the handling of personal data. Examples include the Personal Information Protection and Electronic Documents Act (PIPEDA) in Canada, the Data Protection Act in the United Kingdom, and the Personal Data Protection Act in Singapore. Organizations operating internationally or within particular countries must guarantee compliance with the appropriate data protection regulations applicable to their activities.

5. Data breach notification requirements: Privacy and data protection laws generally place responsibilities on enterprises to inform people and appropriate authorities in the case of a data breach

that constitutes a danger to individuals' rights and freedoms. Prompt notification and adequate procedures to limit the consequences of data breaches are critical for preserving confidence in AI systems.

6. Privacy by design and privacy impact assessments: Privacy by design is a method that advocates the inclusion of privacy and data protection principles into the design and development of AI systems. Privacy impact evaluations assist identify and manage any privacy problems linked with AI initiatives. Implementing these techniques guarantees that privacy issues are entrenched throughout the AI development lifecycle.

7. Cross-border data transfers: When personal data is sent across borders, enterprises must guarantee compliance with appropriate laws and methods for data transfer, such as the EU Standard Contractual Clauses, Binding Corporate Rules, or other recognized mechanisms. These procedures guarantee that personal data is appropriately safeguarded when transmitted to countries lacking an acceptable degree of data protection.

8. permission requirements: Privacy regulations generally require companies to get people' permission for the gathering and processing of their personal data. AI systems must conform to permission rules, ensuring that users offer informed and explicit agreement for their data to be used in AI applications.

Understanding and complying with privacy and data protection rules are critical for firms developing and implementing AI systems. It helps preserve people' privacy rights, create trust with users, and avoid legal and reputational costs associated with non-compliance.

7.5 Ethical Considerations in AI Applications
7.5.1 Healthcare and Medical AI

The use of AI in healthcare and medical applications presents several ethical problems. While AI has the potential to improve patient outcomes and enhance healthcare delivery, it also offers distinct ethical difficulties. Here are some significant ethical issues in healthcare and medical AI:

1. Patient Autonomy: Respecting patient autonomy is vital in AI applications. It is vital to guarantee that patients have the right to make informed choices about their healthcare and comprehend the ramifications of AI-driven suggestions or actions. Transparent communication and collaborative decision-making between healthcare providers and patients are crucial.

2. Trust and Transparency: AI algorithms and systems should be transparent and explainable, allowing healthcare professionals and patients to understand how they arrive at judgments or suggestions. Transparent AI builds confidence and lets people assess the dependability and possible biases of AI systems.

3. Data Privacy and Confidentiality: Protecting patient privacy and protecting the confidentiality of medical data are crucial. Healthcare AI systems must comply to relevant privacy and data protection legislation and adopt rigorous security measures to preserve sensitive patient information.

4. Bias and Fairness: AI systems should be built and trained utilizing varied and representative datasets to reduce biases. Bias in AI may lead to discrepancies in healthcare delivery and outcomes. Regular examination and monitoring of AI systems for fairness and biases are important to provide equal treatment for all patients.

5. Accountability and Liability: Clear lines of accountability and responsibility should be created in AI applications. Healthcare practitioners and organizations must be responsible for the choices made or supported by AI systems. Adequate processes should be in place to address possible damage or mistakes produced by AI systems.

AI Revolution in Medicine, The Future of Healthcare

6. Informed permission: Informed permission is vital when employing AI in healthcare. Patients should be given with explicit information regarding the participation of AI in their treatment, including possible advantages, hazards, and restrictions. Informed permission should also encompass the use of patient data for AI development and research purposes.

7. Human-AI cooperation: Effective cooperation between healthcare practitioners and AI technologies is crucial. AI should be considered as a tool to complement human knowledge rather than replacing healthcare workers. Human monitoring and intervention should be accessible to assure the correctness and appropriateness of AI-driven judgments.

8. Continual review and Monitoring: Regular review and monitoring of AI systems' performance, accuracy, and influence on patient outcomes are vital. This helps detect and correct any biases, mistakes, or unforeseen repercussions that may develop in the usage of AI in healthcare.

9. Equity and Accessibility: AI in healthcare should be developed and implemented in a way that fosters equity and accessibility. Efforts should be taken to ensure that AI serves all patient populations, especially those from varied socioeconomic backgrounds and underrepresented groups, and does not worsen current healthcare gaps.

10. Ethical Research and Development: Ethical issues should govern the research and development of AI in healthcare. Research utilizing AI should comply to ethical principles, such as gaining adequate permission, maintaining privacy and data security, and weighing possible risks and benefits to patients and society.

Addressing these ethical problems is vital for competent and ethical deployment of AI in healthcare and medical applications. It needs cooperation among healthcare practitioners, AI developers, legislators, and regulatory authorities to build rules and frameworks that assure the ethical use of AI technology in the best interest of patients and society.

7.5.2 Autonomous Vehicles

The introduction of autonomous vehicles (AVs) presents serious ethical problems. As self-driving vehicles grow more powerful and common, it is vital to address the ethical problems involved with their deployment. Here are major ethical issues in the context of autonomous vehicles:

1. Safety: Safety is a fundamental ethical problem with autonomous cars. AVs should be built and programmed to emphasize the safety of passengers, pedestrians, and other road users. Balancing the protection of residents with the avoidance of damage to others is a crucial ethical challenge in AV development.

2. Decision-Making in Critical scenarios: AVs may confront critical scenarios when they have to make split-second judgments with possible life-and-death effects. Ethical criteria need to be devised to decide how AVs should react in certain instances, considering variables such as limiting injury, observing traffic regulations, and adhering to legal and moral principles.

3. accountability and Responsibility: Determining accountability and responsibility in the case of accidents or incidents with autonomous cars is an ethical and legal dilemma. Clear standards and regulations are important to divide responsibilities between the vehicle owner, manufacturer, software developer, and other relevant parties.

4. Privacy and Data Security: Autonomous cars acquire large quantities of data, including location information and sensor data. Protecting the privacy and security of sensitive data is vital. Ethical rules should cover how data is acquired, kept, and utilized, assuring

people' permission and protecting against illegal access or exploitation.

5. Transparency and Explainability: The decision-making processes of autonomous cars should be clear and explainable. Users and regulators need to understand how AVs function and the reasons driving their behaviour. This involves openness in the algorithms, data sources, and training methods utilized by AV systems.

6. Social Impact and equitable: AV deployment should evaluate possible social consequences and equitable concerns. Ensuring that AV technology helps all parts of society and does not worsen existing disparities is vital. Access to AVs, job consequences for drivers, and fair distribution of benefits should be examined.

7. Ethical Research and Testing: Ethical issues should govern the research and testing of autonomous vehicles. Safe testing techniques, informed permission of participants, and openness in data collection and analysis are vital. Researchers and developers should follow ethical principles to avoid possible hazards and assure the ethical development and deployment of AV technology.

8. Regulatory Framework: Establishing a thorough regulatory framework is important to handle the ethical problems related with autonomous cars. Governments and regulatory agencies play a critical role in creating standards, guaranteeing compliance, and fostering responsibility in AV development and implementation.

Addressing these ethical problems is vital for the appropriate integration of autonomous cars into society. Collaboration among politicians, industry stakeholders, and ethicists is important to set ethical norms and laws that assure the safe and ethical use of autonomous vehicle technology while recognizing society values and objectives.

7.5.3 Online platforms and social media

Due to the widespread impact that social media and online platforms have on people, communities, and society at large, these ethical issues have received a lot of attention. In the context of social media and online platforms, the following are important ethical considerations:

1. Privacy and data protection: Users' personal information is gathered in massive quantities by social media sites. Privacy and data protection must be guaranteed. The collection, storage, and use of user data should be covered by ethical standards, and users should have access to controls over their data and openness on data practices.

2. User Consent and Autonomy: Social media networks need to make getting users' informed consent for data collection and use a top priority. Users should be able to choose their privacy preferences and turn off targeted advertising so that they can have control over the content they share and receive.

3. Algorithmic Transparency and Bias: Social media networks should utilize transparent, impartial algorithms to curate content, personalize feeds, and offer recommendations. The potential biases in algorithms should be addressed by ethical standards, which should also protect consumers from offensive or harmful content.

4. Safety and content moderation: It is morally required to provide a welcoming and safe online community. For the purpose of tackling concerns like hate speech, harassment, false information, and dangerous content, platforms should have explicit policies for content moderation. A major ethical difficulty is striking a balance between the necessity for responsible content control and the right to free speech.

5. Social media platforms have the potential to have an impact on the mental health and wellbeing of their users. Platforms should be responsibly designed, with aspects that encourage addiction,

comparison, and a poor self-perception minimized. Important ethical goals include promoting digital well-being, encouraging good relationships, and offering mental health support.

6. Social media platforms have a tremendous impact on the dissemination of false information and fake news. To reduce the transmission of false information and advance true information, ethical rules should encourage responsible content sharing, fact-checking systems, and algorithmic interventions.

7. Social media platforms should have clear policies and procedures for handling user complaints, disputes over content, and other user issues. This is known as platform governance and accountability. It is essential to ensure accountability for platform choices and actions. Transparency in platform governance, including decision-making procedures and the inclusion of other perspectives, should also be addressed through ethical principles.

8. Digital Divide and Social Impact: Ethical concerns should take into account how social media platforms affect vulnerable people, as well as how they affect democracy and social cohesion. Important ethical objectives include ensuring equal access to platforms, resolving digital disparities, and limiting the amplifying of injustices.

9. In order to prevent misleading practices and to safeguard vulnerable users, such as children and teenagers, social media platforms should have explicit rules for advertising and influencer marketing. The use of ethical advertising techniques and full disclosure of sponsored content are critical.

The proper and ethical use of social media and internet platforms depends on addressing these ethical issues. Platform providers, policymakers, users, and other stakeholders must work together to create ethical standards, laws, and accountability systems that put the needs of users, their privacy, and a safe online environment first.

AI Revolution in Medicine, The Future of Healthcare

Financial Services and Automated Decision-Making 7.5.4

Algorithmic decision-making is becoming common in the financial services sector, having an impact on things like fraud detection, investment advice, and credit rating. However, there are significant ethical questions raised by the employment of algorithms in the financial sector. In the context of algorithmic decision-making in financial services, the following are important ethical considerations:

1. Fairness and Bias: Financial services algorithms should be developed to ensure fairness and reduce bias. Ethics standards should include the effects of algorithmic decisions on various demographic groups as well as any biases in the data that algorithms employ. To prevent discriminatory outcomes, fairness in lending, insurance, and other financial services should be given top priority.

2. Financial institutions should make an effort to make their algorithmic decision-making processes clear and comprehensible to customers. Users should be given explicit explanations of how algorithms make judgments, along with the considerations taken into account and the justifications for any recommendations or rejections. Transparency promotes trust and enables users to see and correct any prejudices.

3. Data Security and Privacy: It is the duty of financial organizations to ensure the security and privacy of consumer data. Strong data protection procedures, safe data transmission, and gaining users' informed agreement for data use should all be ethical issues. Users should be able to manage their personal information and be informed of how algorithms use it.

4. Accountability and Redress: In situations when algorithmic decisions negatively affect clients, financial institutions should put in place measures for accountability and redress. The handling of disagreements, grievances, and mistakes brought on by algorithmic

decision-making should be included in ethical rules, guaranteeing that those who are harmed have legal options.

5. Financial institutions should ensure the accuracy and resilience of their algorithms. Algorithmic Validation. To reduce errors, avoid malicious manipulation, and preserve reliability, algorithms should undergo thorough testing and validation. It's crucial to continuously monitor and audit algorithms in order to identify biases or unintended effects and fix them.

6. Financial institutions must inform and empower customers about algorithmic decision-making processes, including their limitations and associated hazards. Giving clients knowledge and understanding enables them to make wise financial decisions and promotes openness in the application of algorithms.

7. Algorithms can improve decision-making, but human inspection and intervention are still necessary to achieve moral and responsible outcomes. Financial institutions ought to have systems in place that permit human intervention when it's appropriate, particularly in tricky or delicate situations. The context, moral judgment, and empathy that human specialists may offer may be lacking in algorithms.

8. Regulatory Compliance: With regard to algorithmic decision-making, financial institutions should abide by all applicable laws, rules, and industry standards. To protect the rights and interests of customers, ethical principles should cover compliance requirements, openness in algorithmic practices, and regulatory monitoring.

The responsible and ethical use of algorithms in financial services requires addressing these ethical issues. Establishing moral standards, laws, and procedures that give fairness, transparency, privacy, and accountability a priority in algorithmic decision-making calls for cooperation between financial institutions, authorities, and other interested parties.

7.6 Dealing with Discrimination and Bias in AI Systems

7.6.1 Recognizing Discrimination in AI Algorithms

Artificial intelligence (AI) bias is the existence of unfair or biased outcomes that disproportionately affect particular people or groups. The algorithms in AI systems can reinforce and amplify biases if the historical data they use contains biases or reflects societal imbalances. Understanding the many biases that can appear in AI algorithms is essential:

1. Sampling bias: If the training data used to create AI algorithms is biased or not representative of the population as a whole, bias may result. Skewed outcomes can be caused by biased training data since the algorithms may not have learned from a variety of examples.

2. Selection bias happens when specific data points are purposefully left out of the training dataset. As a result, the algorithm may have learned from only a portion of the available relevant data, which can result in incomplete or biased representations.

3. Algorithmic Bias: If the algorithms themselves have biases built into their design or decision-making processes, algorithmic bias may result. Biases may be introduced by the selection of characteristics, the weighting of particular elements, or the modeling presumptions used when developing the algorithm.

4. Contextual Bias: Biases that appear in particular circumstances or applications are referred to as contextual biases. Data from one population or region may not generalize well to data from other populations or regions, resulting in biased results in many scenarios for AI systems.

5. Proxy Bias: When AI systems use proxy variables that are connected with sensitive traits (such race or gender) but are not included as inputs directly, proxy bias occurs. Even if these sensitive attributes are not included explicitly in the algorithm, the results may nevertheless be discriminatory.

6. Feedback Loop Bias: When biased results from AI systems are given back into the system as new training data, the earlier biases are reinforced and amplified. This may start a cycle of injustice or discrimination that never ends.

A multifaceted strategy comprising data collecting, algorithm design, and continual monitoring is needed to address bias in AI algorithms. Among the tactics to lessen bias are:

1. Data that is Diverse and Representative of the Population: Training data that is diverse and reflects the population being serviced might assist minimize bias. Data from various demographic groups, socioeconomic backgrounds, and geographic areas should be carefully included.

2. Data cleaning and preprocessing procedures can be used to find and reduce biases in a dataset. To prevent bias from being introduced during the preprocessing step, this may entail eliminating outliers, balancing class distributions, and carefully addressing missing data.

3. Algorithmic Fairness: When creating algorithms, fairness principles should be taken into account explicitly. Fairness-aware algorithms can work to prevent favoring or disadvantageating particular people or groups based on delicate characteristics and minimize divergent effects on various groups.

4. Regular Auditing and Evaluation: It's important to continuously monitor and audit AI systems in order to spot biases and correct them. Regular examination of algorithm performance

across various subgroups can aid in the early detection and reduction of bias.

5. Promoting openness and explainability in AI systems can aid in identifying and correcting biases. Stakeholders can better understand and examine any potential biases in the system by receiving detailed explanations of how algorithms generate decisions.

6. Establishing ethical standards and legal frameworks that address bias and discrimination in AI systems can provide developers and users direction and hold them accountable. Best practices, fairness tenets, and penalties for non-compliance can all be outlined in such rules.

The persistent task of eliminating bias and discrimination in AI systems calls for cooperation amongst data scientists, subject matter experts, ethicists, decision-makers, and impacted populations. It is possible to create AI systems that are fair, impartial, and support equitable results for all people and groups by proactively addressing bias and putting in place effective mitigation techniques.

7.6.2 Reducing Bias in Data and Algorithms Used for Training

To provide impartial and fair AI systems, bias must be reduced in training data and algorithms. Here are a few tactics to combat bias at various stages:

1. Data gathering and preparation:

• Ensure diverse and representative datasets: To prevent under- or overrepresenting particular groups, gather data from a variety of sources and demographics.

• Recognize potential biases in the data and correct them during preprocessing by doing a comprehensive examination of the data.

This could entail adjusting class distributions, eliminating or correcting biased data points, or adding more samples to the dataset.

2. Development of Bias-Aware Algorithms:

• Feature engineering and selection: Exercise caution when choosing features for AI algorithms and stay away from sensitive traits that can provide discriminating results. Think about how each attribute affects prejudice and fairness.

• Fairness aims and constraints: Include fairness objectives and limitations in the algorithm design process. Define fairness goals, such as balancing mistake rates between various groups or reducing disparate impact, and then optimize the algorithm in accordance with those goals.

• Consistent algorithm monitoring and auditing: Analyze the algorithm's performance over time to spot and correct biases that might manifest themselves during training or deployment. Audit the algorithm frequently to determine whether it is fair and to eliminate any unintentional biases.

3. Testing and evaluating bias regularly:

• Assess algorithm performance across diverse subgroups: Examine the algorithm's output for various demographic groups to spot any biases or discrepancies. This includes rating performance according to a person's race, gender, age, or other pertinent characteristics.

• Perform fairness assessments: To gauge and assess the fairness of the algorithm's results, use specified fairness criteria. Metrics like equalized odds, statistical parity, or demographic parity can be included in this.

• Integrate user feedback and viewpoints: To comprehend the impact and potential biases of the AI system, gather user and stakeholder feedback. Consider their viewpoints and insights as you work to improve the fairness of the system.

4. Clarity and Transparency:

• Offer justifications for AI decisions: Ensure that AI systems are built to offer justifications or explanations for their choices. This makes it easier for users and other stakeholders to understand how the system arrived at its results and spot any biases.

• Keep thorough records of the algorithm's development, including the choices that were made for feature selection, preprocessing, and model design. This makes it easier to conduct external audits or reviews and identifies sources of prejudice.

5. Ethical Evaluation and Supervision:

• Create interdisciplinary teams or committees to analyze and evaluate the ethical implications of AI systems. Establish ethical review boards or committees. To ensure fairness and reduce bias, these organizations can offer direction and supervision.

• Comply with ethical standards and laws: Obey ethical standards and laws pertaining to bias and justice in the creation and application of AI. Keep up with new guidelines and best practices in the industry.

By putting these tactics into practice, businesses can try to create AI systems that are more impartial, fair, and responsible, hence lowering the possibility of discriminating consequences.

7.6.3 Making AI Systems Fair and Equitable

To prevent maintaining current social prejudices and inequities, it is essential to ensure justice and equity in AI systems. Here are a few crucial things to keep in mind when promoting fairness and equity:

1. Clearly establish fairness standards and objectives that are in line with society ideals and moral precepts. Depending on the situation, many fairness concepts—such as demographic parity, equalized chances, and equal opportunity—might be pertinent. To take into account various viewpoints, think about incorporating various stakeholders in the definition of these criteria.

2. Address biased training data: Recognize biases in training data that could result in unjust outcomes and take appropriate action. To do this, the training dataset must be diverse and representative, and biases in the data gathering procedure must be minimized.

3. Regularly assess AI system performance for biases and disparities in order to spot and address prejudices and inequalities that may exist across certain protected characteristics or demographic groups. Regular audits, impact analyses, and fairness assessments can aid in identifying possible problems and directing adjustments.

4. Improve interpretability and transparency by giving reasons for decisions and actions taken by AI systems. This makes it possible for users and other interested parties to comprehend how the system operates and spot potential biases. This can be accomplished using strategies like interpretable machine learning, model-agnostic explanations, and transparency measures.

5. Work together with a variety of stakeholders to achieve a thorough knowledge of potential biases and their effects, including affected groups, subject matter experts, ethicists, and regulatory agencies. Collaboration may encourage accountability while identifying blind spots and incorporating many viewpoints.

6. Algorithms should be updated and improved frequently in order to address biases and enhance fairness. To iteratively enhance the system's performance, this may entail retraining models with

updated datasets, fine-tuning fairness restrictions, or incorporating user feedback.

7. Conduct bias and fairness audits: Implement systematic audits to evaluate the equity and fairness of AI systems, including audits of algorithms, data, and impacts. An impartial assessment of fairness and equity can be provided by external audits or third-party evaluations.

8. Create rigorous ethical review processes and use them to assess the potential ethical ramifications and biases of AI systems. Ethical review committees or boards can act as watchdogs and guarantee that fairness and equity norms are followed.

9. Regularly inform and prepare stakeholders: Inform developers, data scientists, and decision-makers on bias, fairness, and equity in AI systems. This promotes a culture of fairness and guarantees that everyone involved is aware of their responsibility for advancing equitable AI.

10. Respect for rules and regulations: Obey rules, regulations, and standards that deal with fairness, equity, and non-discrimination in AI. To maintain compliance, keep up with recent legal and regulatory developments in the industry.

Organizations can design AI systems that uphold fairness, equity, and social responsibility by using these tactics, helping to create a society that is more inclusive and equitable.

7.7 Moral Issues with AI Decision-Making
7.7.1 Ethical Dilemmas in Autonomous Systems

Autonomous systems driven by AI bring distinct ethical concerns and difficulties. Here are some significant ethical concerns linked with autonomous systems:

1. Human safety vs. machine decision-making: Autonomous systems, such as self-driving automobiles or autonomous drones, may confront scenarios where they need to make split-second judgments that might effect human safety. Ethical difficulties occur when considering whether the system should prioritize the safety of the user, other persons, or a mix of both in such scenarios.

2. culpability and accountability: Assigning culpability and accountability in circumstances when an autonomous system causes damage or makes a bad choice may be hard. Determining who is accountable for the acts or choices made by the system involves ethical problems connected to legal frameworks and the need for proper methods to resolve responsibility.

3. Transparency and explainability: Autonomous systems frequently involve complicated algorithms and decision-making processes that may not be clearly explainable to humans. The lack of transparency might create issues about the understandability and trustworthiness of the system, especially in important sectors where human lives are at risk.

4. Unforeseen and new scenarios: Autonomous systems may meet situations that were not expressly accounted for during their training or development. These unexpected scenarios might raise ethical issues since the system may not have learnt proper reactions or behaviors, leading to ambiguous or perhaps biased results.

5. Privacy and data usage: Autonomous systems frequently depend on enormous volumes of data for training and decision-making. Ethical difficulties occur when balancing the advantages of data-driven decision-making with the privacy concerns of people whose data is being gathered and utilized by the system.

6. Inherent biases and discrimination: Autonomous systems might unwittingly propagate biases and prejudice contained in the training data. Ethical issues occur when these prejudices lead to

unjust or discriminatory results, notably in sectors such as recruiting, lending, or criminal justice.

7. Ethical trade-offs and decision prioritization: Autonomous systems may confront circumstances where they need to make ethical trade-offs, such as estimating the worth of various human lives or balancing the risks and advantages of alternative courses of action. These judgments include significant ethical issues and need careful design and programming.

8. Human supervision and intervention: Determining the proper amount of human oversight and intervention in autonomous systems is an ethical dilemma. Striking the correct balance between autonomous decision-making and human control is vital to maintain accountability, safety, and ethical decision-making.

Addressing these ethical challenges needs multidisciplinary cooperation including ethicists, AI researchers, politicians, and other stakeholders. Developing rigorous rules, standards, and guidelines may assist guide the design, development, and deployment of autonomous systems, ensuring that they comply to ethical principles and correspond with society norms.

7.7.2 Ethical Considerations in AI-Enabled Decision Support Systems

AI-enabled decision support systems (DSS) provide essential aid to professionals in numerous disciplines, but they also present ethical problems that need to be addressed. Here are some significant ethical issues linked with AI-enabled decision assistance systems:

1. confidence and dependency: Users of AI-enabled DSS may put great confidence and reliance on the system's suggestions or conclusions. Ethical difficulties emerge when the system's

performance or limits are not clearly conveyed, leading to blind confidence or overreliance on its outputs without rigorous review.

2. openness and explainability: Decision support systems generally involve complicated AI algorithms and models, which may lack openness or explainability. Users may find it challenging to grasp the logic behind the system's suggestions or judgments. Ensuring openness and explainability is vital for developing confidence and enabling users to judge the system's dependability and any biases.

3. Accountability and responsibility: AI-enabled DSS may affect key choices that have substantial ramifications for people or organizations. Determining accountability and responsibility in circumstances when the system creates unpleasant consequences or mistakes may be tricky. It is necessary to create clear lines of responsibility and processes for correcting any damage caused by the system's recommendations or actions.

4. Data quality and biases: AI-enabled DSS relies on data for training and decision-making. Ethical difficulties occur if the data used to train the system is of low quality, prejudiced, or reflects society biases. Biases inherent in the data might lead to biased suggestions, increasing existing inequities and injustice.

5. Human supervision and decision-making: While AI-enabled DSS may give helpful insights and suggestions, it is crucial to evaluate the right amount of human oversight and decision-making. Ethical difficulties emerge when considering whether the final choice should exclusively depend on the system's advice or whether human judgment should play a more substantial role.

6. Informed consent and user comprehension: Users of AI-enabled DSS should have a full awareness of the system's capabilities, limits, and possible hazards. Ethical issues include assuring informed permission, giving appropriate information about

how the system works, and addressing any unintended effects or biases that may impair the user's decision-making process.

7. Equity and fairness: AI-enabled DSS should be built and taught to guarantee fairness and equity in their recommendations or judgements. Ethical difficulties occur when the system's results disproportionately benefit or disfavor particular groups, perpetuating existing inequities or biases in the decision-making process.

8. ongoing monitoring and improvement: AI-enabled DSS should undergo ongoing monitoring and assessment to detect and correct any biases, limits, or unexpected effects that may occur over time. Regular updates and modifications are important to guarantee the system stays ethical, accurate, and consistent with the developing demands and beliefs of the users.

Addressing these ethical problems needs coordination among AI developers, domain experts, ethicists, and regulatory agencies. Developing ethical principles, standards, and auditing methods particular to AI-enabled decision support systems may assist guarantee that these systems function with transparency, fairness, accountability, and respect for user autonomy.

7.7.3 Human-AI Collaboration and Responsibility

As AI systems become more integrated into numerous sectors, like healthcare, transportation, and finance, the notion of human-AI cooperation and shared accountability becomes more relevant. Here are some ethical concerns relating to human-AI cooperation and responsibility:

1. Shared decision-making: Human-AI cooperation should stress shared decision-making between humans and AI systems. The idea is to use the skills of both parties to arrive at better solutions. Ethical issues include ensuring that people have the last say in choices and

that AI systems offer clear and intelligible explanations for their suggestions.

2. Human supervision and intervention: While AI systems may give significant insights and suggestions, human oversight and intervention are needed. Humans have the obligation to critically analyze AI-generated outputs, challenge biases or inaccuracies, and make informed judgments based on their knowledge, judgment, and ethical concerns.

3. Training and education: Both humans and AI systems require adequate training and education to support efficient cooperation. Humans should get training on understanding AI capabilities and limits, evaluating AI outputs, and making educated judgments. AI systems should be regularly educated on relevant data and updated to better their performance and fit with human expectations.

4. Accountability for results: In human-AI cooperation, accountability for outcomes should be explicitly stated. It is necessary to create methods for allocating blame when mistakes or injury occur. This involves identifying the roles and responsibilities of people and AI systems, documenting the decision-making process, and providing adequate management and monitoring of AI systems.

5. continual review and improvement: Human-AI cooperation needs continual examination of the AI system's performance, effect, and possible biases. Regular evaluations assist detect and resolve ethical flaws, prejudices, or unexpected effects that may occur. input loops should be built to absorb user input and alter AI systems to increase their performance and ethical alignment.

6. Ethical concerns in AI design: AI systems should be created with ethical issues in mind, such as justice, transparency, and accountability. This involves adding tools to identify and eliminate

biases, offering explanations for AI-generated outputs, and ensuring that the system respects privacy and data protection.

7. Trust and transparency: Trust is important for efficient human-AI cooperation. AI systems should be clear about their capabilities, limits, and decision-making processes. Clear communication channels should be developed to promote conversation and resolve user issues. Trust-building measures should involve openness in data utilization, system performance, and decision-making factors.

8. Value alignment: Human-AI cooperation should accord with human ideals and social objectives. AI systems should be built and deployed in a way that respects human dignity, promotes justice, and resolves social issues. Ethical issues include evaluating the possible influence on vulnerable people, fostering inclusion, and avoiding reinforcement of existing prejudices or injustices.

Overall, human-AI cooperation should be driven by ethical standards, ensuring that AI systems are created and deployed in ways that complement human judgment, respect human values, and increase human well-being. Open conversation, multidisciplinary cooperation, and continuing review are necessary to handle the many ethical difficulties that occur in this collaborative partnership.

7.8 AI Ethics in Research and Development

7.8.1 Ethical Guidelines for AI Research

As AI continues to grow, ethical norms for AI research and development play a key role in assuring responsible and good results. Here are some significant ethical issues and requirements for AI research:

1. Beneficence and non-maleficence: AI research should strive to maximize benefits and reduce damage to people and society. Researchers should examine the possible influence of their study on

diverse stakeholders and prioritize the well-being and safety of users and impacted communities.

2. Fairness and equity: Researchers should attempt to design AI systems that are fair and egalitarian, without introducing unfair discrimination or reinforcing existing prejudices. This entails correcting biases in training data, algorithms, and decision-making processes to provide fair treatment and opportunity for all persons.

3. Privacy and data protection: AI research should conform to standards of privacy and data protection. Researchers should get informed permission from participants, manage personal data securely, and anonymize or de-identify data to respect people' privacy rights. Transparency regarding data gathering and utilization is also vital.

4. openness and explainability: AI research should promote openness and explainability to build understanding and trust. Researchers should aim to make AI systems and their decision-making processes visible and give explanations for the outputs or suggestions provided by AI systems.

5. Accountability and accountability: Researchers should assume responsibility for the creation and deployment of AI systems. This entails completing thorough testing and validation, admitting limits and possible hazards, and resolving any unexpected effects or biases that result from the technology.

6. Collaborative and multidisciplinary approaches: Ethical AI research benefits from cooperation across fields and involvement with various stakeholders. Researchers should seek feedback from experts in relevant disciplines, include impacted groups, and examine the social ramifications of their work throughout the study process.

7. Continuous assessment and improvement: AI research should be a dynamic and iterative process. Researchers should regularly analyze the ethical implications of their work, monitor the effect of AI systems, and take efforts to reduce any bad outcomes. Feedback methods and processes for user feedback and redress should be implemented.

8. Responsible publishing and sharing of results: Researchers should publish their findings in a transparent and responsible way, revealing both good and negative outcomes. This involves reporting any conflicts of interest, giving adequate information for replication, and avoiding excessive statements that might mislead or generate unreasonable expectations.

9. Ethical evaluation and supervision: Ethical issues in AI research should be subject to thorough examination and oversight. Researchers should adhere to institutional and professional ethical principles, get essential permits for human subjects research, and guarantee compliance with applicable laws and regulations.

10. Global viewpoints and cultural sensitivity: AI research should incorporate global perspectives and cultural diversity. Researchers should realize that ethical standards and values may vary across various civilizations and circumstances. Efforts should be taken to eliminate cultural prejudice and interact with varied viewpoints to guarantee the wide application and adoption of AI technology.

These ethical standards give a framework to support responsible AI research and development, considering the social effect and consequences of AI systems. Adhering to these rules helps promote trust, reduce risks, and guarantee that AI technologies are developed and implemented in a way that complies with ethical principles and societal norms.

7.8.2 Responsible AI Innovation and Deployment

Responsible AI innovation and deployment are critical to guarantee that AI technologies are created, implemented, and utilized in a way that complies with ethical concerns and social values. Here are some crucial elements to consider for ethical AI innovation and deployment:

1. Human-centered design: AI systems should be created with a focus on human needs and values. It is crucial to incorporate end-users and stakeholders throughout the development process to understand their needs, concerns, and possible effect. Human-centered design principles assist construct AI systems that are intuitive, useful, and solve real-world challenges successfully.

2. Ethical concerns from the outset: Ethical issues should be included into the AI innovation process from the beginning. Researchers and developers should proactively identify and address ethical concerns, possible biases, and unexpected effects of AI systems. This requires doing rigorous risk assessments, assessing the social effect, and obtaining opinion from ethical experts.

3. Testing and validation: Rigorous testing and validation are necessary before implementing AI systems. This involves assessing the system's correctness, dependability, and resilience across varied data sets and situations. Testing should also concentrate on possible biases, fairness, and the system's performance under real-world settings. Ongoing monitoring and assessment are important to verify the system's continuing efficacy and adherence to ethical norms.

4. Explainability and transparency: AI systems should give explanations for their decision-making processes. This encourages openness and helps establish confidence among end-users, stakeholders, and regulators. Users should have access to intelligible and relevant explanations for how AI systems arrived at their suggestions or findings. Transparent documentation of AI models,

algorithms, and data sources is also vital for external inspection and accountability.

5. Risk management and mitigation: Organizations should implement comprehensive risk management techniques for AI adoption. This entails recognizing possible hazards and devising mitigation methods to mitigate them. Organizations should also have contingency plans in place to address system breakdowns, data breaches, or other unanticipated occurrences. Regular audits and monitoring procedures assist assure continuing compliance with ethical and legal standards.

6. Compliance with legal and regulatory frameworks: AI innovation and deployment should conform with relevant laws and regulations controlling privacy, data protection, fairness, and non-discrimination. Organizations should be knowledgeable about the legal and regulatory environment and actively seek to remedy any legal or ethical deficiencies. Engaging with regulatory bodies and getting legal guidance may assist assure compliance and appropriate deployment.

7. User empowerment and consent: Users should have control over their data and how it is utilized in AI systems. Organizations should give clear information about data collection, storage, and use procedures to acquire informed permission from users. Empowering users with the flexibility to set choices, opt-out, or seek modifications helps respect individual autonomy and privacy.

8. Continuous monitoring and improvement: AI systems should be regularly monitored and analyzed to assess their performance, eliminate biases or unfairness, and discover areas for improvement. Regular input from users and stakeholders helps detect possible ethical concerns or unexpected repercussions. Organizations should have processes in place to absorb input, make appropriate

modifications, and constantly enhance the AI system's ethical and social effect.

9. Collaboration and shared responsibility: Responsible AI deployment needs collaboration among multiple stakeholders, including academics, developers, policymakers, ethicists, and end-users. Open discourse, information exchange, and cooperation help build a common awareness of ethical dilemmas and encourage collective responsibility in resolving them. Partnerships between academia, industry, and regulatory agencies may support responsible AI innovation and implementation.

10. Ethical governance and oversight: Organizations should build internal procedures and structures to guarantee ethical governance and supervision of AI deployment. This may require forming specialized ethics committees, conducting frequent ethical effect assessments, and setting explicit norms and procedures for AI research and deployment. External audits and certifications from independent organizations may further strengthen openness and accountability.

By implementing these principles, businesses may support responsible AI research and deployment, ensuring that AI technologies contribute positively to society while limiting possible risks and damages. Responsible AI practices not only preserve human rights and values but also encourage trust, acceptability, and the long-term viability of AI applications.

7.8.3 Ethical Considerations in Data Collection and Usage

Data gathering and consumption are essential components of AI development and implementation. Ethical issues around data play a significant part in assuring the responsible and ethical usage of AI technology. Here are some significant ethical issues in data collecting and usage:

1. Informed consent: Obtaining informed permission is vital when gathering personal data for AI applications. Individuals should be fully informed about the goal of data collection, the categories of data being collected, how the data will be used, and any possible dangers or ramifications. Consent should be voluntary, explicit, and based on a clear knowledge of the ramifications of data sharing.

2. private protection: Respecting people' private rights is of highest significance in data collecting and utilization. Organizations should comply to privacy laws and regulations and use rigorous security measures to preserve gathered data. Anonymization and encryption measures should be used to safeguard personally identifying information and limit the danger of re-identification.

3. Data transparency and fairness: Transparent data collecting procedures assist establish confidence and assure justice. Organizations should clearly define the data gathering process, the sources of data, and the criteria for data inclusion or exclusion. It is vital to prevent biased data selection or sample practices that may disproportionately harm particular groups or lead to discriminatory conclusions.

4. Data quality and accuracy: Ensuring data quality and correctness is critical to ensure the integrity and dependability of AI systems. Organizations should create methods to check and verify the quality and completeness of acquired data. Data cleaning strategies, including eliminating outliers and correcting data inconsistencies, should be applied to increase data quality.

5. Minimization of data gathering: Data collection should be restricted to what is essential for the targeted AI application. Organizations should avoid collecting excessive or unnecessary data that may create privacy issues or increase the risk of illegal access. Adopting data reduction principles helps eliminate privacy threats and guarantees appropriate data processing.

6. Data ownership and control: Individuals should have control over their data and the opportunity to select how it is used. Organizations should give explicit alternatives for people to view, amend, delete, or limit the use of their data. Respecting data ownership rights and encouraging people to exert control over their data promotes openness and fosters confidence.

7. Avoidance of discriminatory practices: Care should be made to avoid the use of data that may perpetuate or reinforce existing prejudices or discrimination. prejudiced or discriminating data may lead to prejudiced AI systems, aggravating societal disparities. Organizations should employ ways to uncover and reduce biases in data collection, preparation, and algorithmic decision-making.

8. Data sharing and collaboration: Responsible data sharing standards should be followed when cooperating with other organizations or researchers. Data sharing should conform to privacy restrictions, data protection legislation, and ethical norms. Sharing data in a regulated and safe way improves scientific advancement, enables validation of AI models, and assures responsibility.

9. Data governance and accountability: Organizations should have explicit data governance frameworks and methods to guarantee accountability in data gathering and utilization. This involves defining roles and duties, creating data protection rules, and implementing internal controls for data management. Regular audits and evaluations assist monitor compliance and identify opportunities for improvement.

10. Continuous monitoring and review: Ethical concerns in data collecting and utilization should be a continuous effort. Organizations should regularly monitor and analyze their data practices, ensuring conformity with growing ethical standards, regulatory constraints, and public expectations. Regular audits and

evaluations assist uncover possible risks or concerns and allow quick remedial measures.

By preserving these ethical principles in data collection and utilization, businesses may support responsible and ethical AI research and deployment, defend human privacy and rights, and encourage trust and openness with people and communities.

7.9 Ensuring Ethical and Legal Compliance in AI
7.9.1 Ethical Codes and Standards

Ethical ethics and standards play a significant role in guiding the appropriate development and deployment of AI systems. They offer a framework for organizations and practitioners to adhere to ethical principles and best practices. Here are several significant ethical principles and standards in the area of AI:

1. Ethical rules for AI: Various organizations and institutes have produced ethical rules expressly for AI. For example, the European Commission's Ethics Guidelines for Trustworthy AI emphasize fundamental values such as openness, justice, accountability, and human agency. These standards serve as a reference for ethical AI development and implementation.

2. Professional Codes of Conduct: Professional groups and organizations frequently have codes of conduct that describe ethical duties for practitioners working in AI. These norms outline expectations for professional conduct, including concerns relating to privacy, prejudice, fairness, and openness. Practitioners are urged to follow these guidelines to guarantee ethical compliance.

3. Industry Standards: Some sectors have set standards and best practices relevant to AI development and implementation. These standards cover technological elements, data processing, security, and privacy concerns. Adhering to industry standards helps firms achieve compliance with ethical and regulatory obligations.

4. Legal and Regulatory Frameworks: Compliance with legal and regulatory frameworks is vital for ethical AI activities. Governments and regulatory agencies are rapidly adopting legislation and regulations relevant to AI, addressing themes such as data protection, privacy, fairness, and transparency. Organizations should understand and comply with relevant rules and regulations to achieve ethical and legal compliance.

5. Ethical Review Boards: In some circumstances, ethical review boards or committees may be created to examine AI initiatives and guarantee adherence to ethical norms. These review boards analyze the possible dangers, advantages, and ethical implications of AI initiatives, giving advice and suggestions for ethical compliance.

6. International Guidelines and Declarations: International organizations, like as UNESCO and the United Nations, have produced guidelines and declarations that address ethical issues in AI. For example, the Universal Declaration on the Ethics of Artificial Intelligence highlights the relevance of human rights, justice, and responsibility in AI research and deployment.

7. Collaboration and Knowledge Sharing: Collaboration among scholars, practitioners, and policymakers is vital for defining and revising ethical norms and standards. Sharing information and experiences helps create ethical frameworks and encourages responsible AI techniques across diverse domains and industries.

It is crucial for businesses and practitioners to acquaint themselves with appropriate ethical norms and standards and incorporate them into their AI development and deployment processes. Adhering to these ethics and standards not only assures ethical and responsible AI practices but also helps increase public trust and confidence in AI technology. Regular evaluation and revisions to rules and standards are important to keep pace with the growing world of AI and handle new ethical problems.

7.9.2 Regulatory Compliance and Certification

In addition to ethical norms and standards, regulatory compliance and certification procedures play a significant role in guaranteeing the ethical and legal usage of AI technology. These techniques offer a framework for reviewing and verifying that AI systems satisfy particular standards and conform to specified rules. Here are some major components of regulatory compliance and certification in the context of AI:

1. Data Protection and Privacy rules: Many nations and areas have data protection and privacy rules in place to secure people' personal data. Examples include the European Union's General Data Protection Regulation (GDPR) and the California Consumer Privacy Act (CCPA). Organizations working with AI technology need to comply with these requirements, ensuring that personal data is acquired, kept, and processed in a safe and privacy-conscious way.

2. Fairness and Non-Discrimination: AI systems shall not perpetuate prejudices or discriminate against people or groups based on protected traits such as race, gender, or age. Regulations and rules focused on fairness and non-discrimination, such as the Equality Act in the United Kingdom, assist guarantee that AI technology do not worsen existing societal inequities. Compliance with these requirements entails cautious data management, algorithmic openness, and constant monitoring and auditing of AI systems.

3. Medical Device rules: In healthcare settings, AI-enabled medical equipment and software may need to comply with certain rules, such as the U.S. Food and Drug Administration's (FDA) rules regarding medical devices. These laws outline the standards for safety, efficacy, and quality management systems for AI-driven medical devices.

4. Algorithmic openness and Explainability: Some legislative frameworks, such as the EU's GDPR and the planned EU Artificial Intelligence Act, highlight the necessity of algorithmic openness and explainability. These standards oblige enterprises to give people with explicit information about the rationale, relevance, and possible repercussions of automated decision-making processes.

5. Certification Programs: Regulatory agencies and industry groups may provide certification programs that test and certify AI systems for conformity with certain standards and requirements. These certification programs enable firms show their commitment to ethical and responsible AI practices and give confidence to stakeholders, including consumers, regulators, and the public.

6. Regulatory Sandboxes and Pilot Programs: Some regulatory agencies construct sandboxes or pilot programs to aid the testing and implementation of AI technology in regulated circumstances. These programs enable firms to cooperate with authorities, obtain assistance, and maintain compliance with applicable legislation while driving innovation and learning.

Organizations should be knowledgeable about the regulatory environment in their particular areas and proactively interact with regulatory authorities to guarantee compliance with relevant legislation. It is vital to incorporate compliance activities throughout the AI development lifecycle, including data collecting, algorithm development, system deployment, and continuing monitoring. By following to regulatory regulations and getting relevant certifications, firms may show their commitment to ethical and legal AI practices while minimizing risks and assuring responsibility.

7.9.3 Ethical Auditing and Evaluation

Ethical auditing and assessment are crucial techniques for reviewing and assuring the ethical and responsible usage of AI technology. These methods entail methodically assessing and analyzing the ethical implications of AI systems, identifying possible dangers and problems, and applying suitable steps to resolve them. Here are major issues of ethical auditing and assessment in the context of AI:

1. Ethical Frameworks and rules: Organizations might construct ethical frameworks and rules related to AI development and implementation. These frameworks explain the ideals, values, and ethical concerns that underlie the organization's AI projects. Ethical auditing entails analyzing whether AI systems conform with these recognized principles and norms.

2. Impact Assessment: Ethical impact assessments entail analyzing the possible ethical consequences of AI technology throughout their lifespan. This examination analyzes criteria such as justice, openness, accountability, privacy, and social effect. It helps uncover possible biases, prejudice, and unexpected repercussions of AI systems and allows the enterprise to mitigate them.

3. Stakeholder Engagement: Ethical auditing should entail interacting with diverse stakeholders, including users, affected communities, experts, and regulatory organizations. Gathering varied viewpoints helps uncover possible ethical difficulties and ensures that the interests and values of different stakeholders are addressed in the development and deployment of AI systems.

4. Data and Algorithmic Analysis: Ethical auditing entails reviewing the data used to train AI systems and determining if it is representative, impartial, and ethically supplied. It also requires assessing the algorithms and models applied to uncover any possible biases or discriminatory effects. Transparency and explainability of algorithms are crucial factors throughout this procedure.

5. Ethical Decision-Making Processes: Ethical auditing investigates the decision-making processes incorporated inside AI systems. It analyzes if the decision-making conforms with ethical

criteria, fairness principles, and legal obligations. This involves analyzing the extent of human supervision, the capacity to overrule or question system choices, and the processes in place to resolve mistakes or unexpected effects.

6. continuing Monitoring and assessment: Ethical auditing is not a one-time activity but involves continuing monitoring and assessment of AI systems. Regular evaluations let firms detect new ethical challenges, react to changing situations, and constantly enhance the ethical performance of AI systems.

7. Independent Auditing: In certain situations, companies may hire independent auditors or external specialists to undertake ethical audits. Independent audits give an independent examination of AI systems and provide an extra layer of confidence that ethical issues are being effectively handled.

Ethical auditing and assessment should be included into the governance and oversight procedures of AI development and deployment. It helps firms identify and minimize ethical risks, preserve public confidence, and ensure that AI technologies match with ethical values and regulatory requirements. By completing complete ethical audits, firms may show their commitment to responsible and accountable AI processes while fostering transparency and social well-being.

CONCLUSION

In conclusion, AI has the potential to change healthcare, presenting unparalleled chances for enhancing patient care, medical imaging, electronic health records, therapy selection, and more. However, the integration of AI in healthcare also comes with several ethical, legal, and regulatory problems that need to be properly handled.

Throughout this book, we have covered several elements of AI in healthcare, starting with its basics and applications in personalized medicine, medical imaging, and electronic health records. We have covered the revolutionary influence of AI on patient care, including remote monitoring, predictive analytics, pharmaceutical management, patient involvement, mental health, and end-of-life care.

Additionally, we have dug into the ethical aspects surrounding AI-driven healthcare, such as data protection and security, bias and fairness, informed consent, and transparency. We have also studied the regulatory constraints and adoption obstacles that need to be solved to enable the safe and successful integration of AI in clinical practice.

Furthermore, we have studied the future views and possible influence of AI in personalized medicine, targeted medicines, clinical trials, evidence production, patient empowerment, and shared decision-making.

Lastly, we have investigated the function of AI in medical imaging, including its applications in diverse modalities such as ultrasound, computed tomography (CT), magnetic resonance imaging (MRI), nuclear medicine, and molecular imaging. We have reviewed the improvements in AI algorithms for image analysis, segmentation, measurement, lesion identification, characterisation, and functional imaging analysis.

Throughout the book, we have stressed the significance of addressing ethical and legal consequences in AI research and implementation. We have underlined the need for justice, transparency, privacy protection, accountability, and tackling biases and prejudice in AI systems. We have also stressed the necessity of regulatory frameworks, standards, and certification procedures to guarantee compliance and appropriate usage of AI technology in healthcare.

In summary, AI has the potential to change healthcare and enhance patient outcomes. However, it is necessary to overcome the ethical, legal, and regulatory difficulties connected with AI integration to guarantee its responsible and useful usage in healthcare settings. By addressing these factors, we can harness the full potential of AI to enhance healthcare and eventually improve the lives of people globally.

www.ingramcontent.com/pod-product-compliance
Lightning Source LLC
Chambersburg PA
CBHW071448220526
45472CB00003B/721

www.ingramcontent.com/pod-product-compliance
Lightning Source LLC
Chambersburg PA
CBHW071519220526
45472CB00003B/1081